THE NEW SCIENCE OF ASTROBIOLOGY

Cellular Origin and Life in Extreme Habitats

Volume 3

The New Science of Astrobiology

From Genesis of the Living Cell to Evolution of Intelligent Behaviour in the Universe

by

Julian Chela-Flores

KLUWER ACADEMIC PUBLISHERS
DORDRECHT / BOSTON / LONDON

A C.I.P. Catalogue record for this book is available from the Library of Congress.

ISBN 0-7923-7125-9

Published by Kluwer Academic Publishers,
P.O. Box 17, 3300 AA Dordrecht, The Netherlands.

Sold and distributed in North, Central and South America
by Kluwer Academic Publishers,
101 Philip Drive, Norwell, MA 02061, U.S.A.

In all other countries, sold and distributed
by Kluwer Academic Publishers,
P.O. Box 322, 3300 AH Dordrecht, The Netherlands.

COVER: The cupola in the West Atrium of St. Mark's Basilica in Venice representing the biblical interpretation of Genesis, cf., also Chapter 1, p. 14. (With kind permission of the Procuratoria of St. Mark's Basilica.)

Printed on acid-free paper

Printed in the Netherlands.

For Sarah Catherine

Contents

BOOK 1:

ORIGIN OF LIFE IN THE UNIVERSE

INTRODUCTION: **What is astrobiology?**

PART I

CHEMICAL EVOLUTION: FOUNDATIONS FOR THE STUDY OF THE ORIGIN OF LIFE IN THE UNIVERSE

CHAPTER 1. **From cosmic to chemical evolution**

CHAPTER 2: **From chemical to prebiotic evolution**

CHAPTER 3: **Sources for life's origins:
A search for biogenic elements**

PART II

PREBIOTIC EVOLUTION: THE BIRTH OF BIOMOLECULES

CHAPTER 4: **From prebiotic evolution to single cells**

BOOK 2:

EVOLUTION OF LIFE IN THE UNIVERSE

CHAPTER 5: **From the age of prokaryotes to the origin of eukaryotes**

CHAPTER 6: **Eukaryogenesis and evolution of intelligent behavior**

BOOK 3:

DISTRIBUTION AND DESTINY OF LIFE IN THE UNIVERSE

PART I

EXOBIOLOGY OF THE SOLAR SYSTEM: SCIENTIFIC BASES FOR THE STUDY OF THE LIFE OF OTHER WORLDS

CHAPTER 7: On the possibility of biological evolution on Mars

PART II

BIOASTRONOMY: THE STUDY OF ASTRONOMICAL PHENOMENA RELATED TO LIFE

CHAPTER 10: **How different would life be elsewhere?**

CHAPTER 11: **The search for evolution of extraterrestrial intelligent behavior**

CHAPTER 12: **Is the evolution of intelligent behavior universal?**

PART III.

CULTURAL FOUNDATIONS FOR THE DISCUSSION OF THE DESTINY OF LIFE IN THE UNIVERSE

CHAPTER 13: **Deeper implications of the search for extraterrestrial life**

BOOK 4:

SUPPLEMENT

"It's a small world after all"

Foreword

Joseph Seckbach
Editor of the COLE Book Series

The present volume entitled **"The New Science of Astrobiology"** is a continuation of the book series entitled "**Cellular Origin and Life in Extreme Habitats**" (**COLE**), which I have edited and published with Kluwer Academic Publishers, Dordrecht, The Netherlands. The other volumes are: **Enigmatic Microorganisms and Life in Extreme Environments** (1999); **Journey to Diverse Microbial Worlds** (2000), which follow the first book **Evolutionary Pathways and Enigmatic Algae** (1994). The other books of this series (**Symbioses** as well as **Origins**) are in preparation.

Astrobiology is a new area of interdisciplinary studies (this concept has recently been emphasized by NASA). It has been also been referred to as Bioastronomy, or even by the older term of Exobiology; likewise Cosmobiology was sometimes used by the Russian scientific community. The Astrobiological umbrella covers the origin, distribution of terrestrial microorganisms, micro-fossilized records and destiny of life in the Universe. It deals with the discovery of new planets or satellites and the search for extraterrestrial life. Some of these subjects have been extensively discussed in the first two volumes of this series. Terrestrial extremophiles living in severe environments may serve as analogues for life on other worlds possessing the "ingredients" for origin and living conditions of microbes. The recent and future space missions may shine new light on the origin and evolution of life on Earth, as well as on the distribution of life in the Cosmos, including the long time survival of microorganisms in space environments, which are normally challenged by extremophilic conditions.

This new book deals with subjects ranging **"From genesis of the living cell to evolution of intelligent behavior in the universe"**. To our knowledge this is the first book by a single author which deals with the new subject of Astrobiology. This topic of Astrobiology and the frontiers of life are currently the cutting edges of scientific inquiry. Following the Introduction and the Preface by the author comes the text itself. In addition the author has added a glossary to explain many concepts for the non-expert in this field. There are two lists of indices, covering the subjects and the various names of scientists which appear in this book. The list of abbreviations makes the reading easier. Throughout the text there are various relevant illustrations, such as tables and images.

Several international conferences, workshops, congresses have been taking place in recent years; most of them are followed by their corresponding proceedings. The field of Astrobiology is no exception. Recently, there has been an ever increasing plethora of articles covering the origin of life and the possibility of extraterrestrial of life. New books have been mushrooming lately in the various publishing houses around these subjects. Browsing Internet would reveal several sites with articles, books and other sources on Astrobiology, each taking a different approach in dealing with this topic.

The author, Prof. J. Chela-Flores is an original physicist, who has wide-ranging interests in many subjects of the life sciences, as well as other interdisciplinary subjects. Lately his main project of research is the evolution of life and intelligent behavior. Most of his edited books on Astrobiology have been proceedings of conferences which he has organized in Trieste, Italy. These books have increased our insight into this new field. His edited book **"Astrobiology: Origins from the Big-Bang to Civilization"** (published with two other coeditors) is the proceedings of the Iberoamerican School of Astrobiology (which took place in Caracas, Venezuela). This volume provides further insights and complements the above-mentioned books, which have been edited by Chela-Flores and his coeditors (cf., list on page 251 of this volume). It also may be grouped with the new volumes of this series (Seckbach, 1999, 2000), which deal with closely associated subjects.

The editor of this volume wishes to acknowledge and thank the reviewers (of an earlier version of the manuscript), whose constructive comments led to improvements of the present version. Among them are

Professors André Brack (Orleans, France), Frank Drake (SETI Institute, Mountain View, CA), Antonio Lazcano (Mexico City), François Raulin (University of Paris) and Martino Rizzotti (University of Padova, Italy).

References

Seckbach, J. (ed.) (1999) *Enigmatic Microorganisms and Life in Extreme Environments*, Kluwer Academic Publishers, Dordrecht, The Netherlands. http://www.wkap.nl/bookcc.htm/0-7923-5492-3.

Seckbach, J. (ed.) (2000) *Journey to Diverse Microbial Worlds: Adaptation to Exotic Environments*, Kluwer Academic Publishers, Dordrecht, The Netherlands). http://www.wkap.nl/bookcc.htm/0-7923-6020-6

Hebrew University Jerusalem, Israel   **Joseph Seckbach**
seckbach@huji.ac.il
Monday April 16, 2001

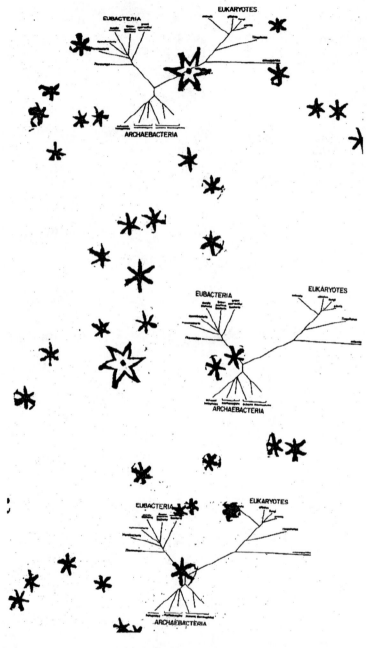

On an original sketch of the many visible stars in the Orion Constellation by Galileo, which appeared in *Sidereus Nuncius*[1], we have superimposed the tree of life that has evolved on Earth. In the case of the evolution of life elsewhere in the universe, a problem that faces the new science of astrobiology is whether the domains of life are universal in such a 'forest of life'.

Preface

Late in the 20th century molecular biology forced upon us the idea that our tree of life, which is technically referred to as a 'phylogenetic tree', has only three 'branches', or domains. These three highest groupings, or taxons, of all life on Earth encompass the three main blueprints of cellular life. As we shall see in detail in Chapter 6, science has now abandoned the taxonomy that retained traditional kingdoms as the highest groupings in the classification of life. The present focus of taxonomy is thus on microorganisms, rather than on multicellular organisms. The traditional classification that survived until the birth of molecular biology (an event which occurred in the middle of last century) had raised animals and plants (instead of microorganisms) to the highest ranking of life on Earth.

In the present book we shall entertain the idea that sooner, or later, we shall be confronted with a variety of trees of life. Indeed, in the following chapters we shall try to persuade the reader that due to the considerable progress in solar system exploration by the main space agencies, sooner rather than later, we shall reach one of the ultimate objectives of science, namely, the identification and subsequent study, not of a single tree of life, but rather of a whole 'forest of life'.

The underlying subject discussed in this book is the new science of astrobiology. It is a very broad interdisciplinary field covering the origin, evolution, distribution and destiny of life in the universe, as well as the design and implementation of missions for the specific purpose of solar system exploration. Our selection of topics in this book has been dictated exclusively by their relevance to these four aspects of astrobiology. Consequently, in view of the vast panorama dealt with, in the last section we have provided a variety of pedagogical additions to the text ("Book 4: Supplement"). The reader is strongly encouraged to take full advantage of this essential part of the text: The glossary contains over 160 entries; by highlighting in italics certain keywords used in the definition of some of the terms, we are referring the readers to other entries of the glossary, where the words in italics are explained.

We indicate with an asterisk the presence of the accompanying entry in the glossary. The name index is also an important part of Book 4, particularly because it contains the names of many scientists and humanists.

A set of notes corresponding to each chapter is an essential part for reading the main text. It helps the reader by making frequent references to the glossary, as well as to the scientific literature. The 48 tables provided represent highly condensed 'windows' into interdisciplinary research ranging from the basic sciences of physics, astronomy and chemistry to the earth and life sciences. Finally, the reader will find a supplementary bibliography for further reading into the various topics dealt with in each chapter.

These additions to the text should make *The New Science of Astrobiology* wholly accessible to both scientists and the interested layperson alike. Nevertheless, due to the interdisciplinary nature of the subject all readers are necessarily expected to invest some time in making full use of all the supplements. In particular, we call the reader's attention to the list of abbreviations that we have provided since, for instance, geologic time is normally cited in the abbreviated form (Gyr BP), instead of the standard English usage (a thousand million, or a billion years before the present). This is compatible with the large time scale of the geologic evolution of the Earth. Without this warning the average reader might have found such notation puzzling.

We feel that the effort involved in following these recommendations is worthwhile. In doing so all of us, scientists, philosophers, the interested theologian and the general reader will be making an effort that will allow them to participate in one of the most interdisciplinary and rewarding activities in the whole of contemporary culture. This comes at a time when revolutionary changes are taking place, and will continue to occur, in our perception of the universe and in our beliefs. We maintain that a constructive dialogue, rather than a destructive debate, between science, philosophy and religion is necessary to integrate harmonically the vast amount of information that is covered by the subject of astrobiology, as understood in the wider sense of the definition given above.

Why is there a need for a new book in this well reviewed subject, with an ever-increasing number of good books being made available?

I felt that a review covering the complete spectrum of astrobiology is still missing at a level accessible to the non-specialist, although a certain effort will be required of him, as we mentioned above. Our small book is a modest attempt to touch on, and integrate, the deeper implications arising from the inclusion of the destiny of life in the universe. This attempt to be comprehensive in reviewing the science of astrobiology comes at the end of a discussion of the first three aspects of our subject:

Indeed, we have started in Book 1 with the origin of life, in the same context as first introduced by organic chemists about 80 years ago. This is followed in Book 2 by some implications of Darwinian evolution of terrestrial biota, a topic that has been extensively discussed by scientists, philosophers and theologians for over a century and a half. Finally, in Book 3 we have gathered together a few aspects of the most exciting, but still not firmly established part of astrobiology, namely the distribution and destiny of life in the universe.

By pointing out the few facts that are known to science, we can only expose the large gap in our current understanding of our own corner of the universe, and ourselves. At the end of Book 3, after integrating two small chapters that touch on two major subjects of our culture, philosophy and theology, we have attempted to present some difficulties that must eventually be faced in the future, when all aspects of our culture are used to consider the major questions that have been raised by astrobiology. By mentioning only one of them: *Are we alone?* perhaps the oldest question raised since classical times, we soon encounter disagreement. In the book of the paleontologist Peter Ward and the astronomer Donald Brownlee *Rare Earth* (Copernicus/Springer, 2000), it is argued, in agreement with what is explained in Book 3, that although microscopic life may be ubiquitous in the universe, the question of the repeatability of human-level of intelligent behavior still remains controversial. On the other hand, we will argue in favor of the ubiquity of human level of intelligence in the cosmos; current ideas of Darwinian evolution militate in its favor, namely, *constrained contingency* (Book 2, Chapter 5), and *convergent evolution* (Book 3, Chapters 12 and 13).

The careful reader may conclude that a proper integration of basic biology into the mainstream of astrobiology is still in its infancy. We would agree. Indeed, we attempt to illustrate in this book that the time is mature for putting the main emphasis of astrobiology on biology rather than on the space and earth sciences.

The present work would not have been written without the formidable combined influence of Abdus Salam and Cyril Ponnamperuma (cf., Fig. P1), who were brought together while they worked on the question of the origin of life in the winter of 1990-1991. Earlier, Abdus Salam had accepted my invitation to visit Caracas, while Simon Bolivar University was celebrating its 10th anniversary in January, 1980; he had been awarded the Nobel Prize in Physics the previous month. We held many discussions on molecular biology during my subsequent sabbatical leave at the International Centre for Theoretical Physics in 1986.

Figure P1: Nobel Laureate Abdus Salam (first from the right) and Cyril Ponnamperuma (Conference Director, first from the left) during the Opening Ceremony of the first Trieste Conference, 26 October, 1992 (courtesy of ICTP Photo Archives).

However, in February 1991 we converged again when Salam introduced me to Cyril Ponnamperuma, suggesting that the conference they had planned, but did not take place in early 1991, should be directed by Ponnamperuma and organized with my collaboration (cf., Fig. P2).

The work that is discussed below has been shaped by the regular meetings of many scientists; some 450 participants had attended the Trieste conferences by 2000, in order to discuss subjects ranging from chemical evolution to the origin, evolution, distribution and destiny of life in the universe. These conferences have taken place regularly from 1992 till 2000 (cf., Figs. P3-P7).

These events have seen the participation, not only of Salam (in 1992) and Ponnamperuma, who directed the first three conferences (1992-1994), but also pioneers in the origin-of-life studies, Sidney Fox, Stanley Miller and Joan Oró. Others that were present were:

Gustav Arrhenius, Vladik Avetisov, Bakasov, Harrick and Marghareta Baltscheffsky, Laurence Barron, Francesco Bertola, André Brack, Graham Cairns-Smith,

Giorgio Careri, Mohindra Chadha, David Cline, Marcello Coradini, John Corliss, Cristiano Cosmovici, George Coyne, Paul Davies, Donald De Vincenzi, Klaus Dose, Frank Drake,

Figure P2: Partial view of the participants of the first Trieste Conference. Salam is sixth from the left to right and Ponnamperuma is seventh (1992 courtesy of ICTP Photo Archives).

Giancarlo Genta, Georgi Gladyshev, Vitali Goldanskii, J. Mayo Greenberg, Kaoru Harada, Jean Heidmann, Gerda Horneck, Yoji Ishikawa, Jamal Islam, Michael Ivanov, Otto Kandler, Lajos Keszthelyi, Richard D. Keynes, Kensei Kobayashi, Mikhail Kritsky, Igor Kulaev,

Figure P3: Participants of the second Trieste Conference (1993, courtesy of ICTP Photo Archives).

Narendra Kumar, Doron Lancet, Antonio Lazcano, Guillermo Lemarchand, Alexandra MacDermott, Claudio Maccone, Koichiro Matsuno, Clifford Matthews, Michel Mayor,

Figure P4: Participants of the third Trieste Conference (1994, courtesy of ICTP Photo Archives).

Christopher McKay, David McKay, Stephen Moorbath, Rafael Navarro-González, Alicia Negrón-Mendoza, Marc Ollivier, Tahiro Oshima, Tobias Owen, Cynthia Phillips, Daniel Prieur, François Raulin, Martino Rizzotti, Robert John Russell, Takeshi Saito, Manfred

Figure P5: Participants of the fourth Trieste Conference (1995, courtesy of ICTP Photo Archives).

Schidlowski, J. William Schopf, Jean Schneider, Peter Schuster, Joseph Seckbach, Everett Shock, Jill Tarter, Wang Wenqing, Frances Westall and Yu-Fen Zhao.

Figure P6: Participants of the fifth Trieste Conference (1997, courtesy of ICTP Photo Archives).

Figure P7: Participants of the sixth Trieste Conference [1] (2000, courtesy of ICTP Photo Archives and Massimo Silvano).

I would like to underline and, at the same time, to express my special gratitude to the ICTP Director Miguel Angel Virasoro, for not only continuing to encourage the Trieste series of conferences since 1995, but also for allowing us to expand these activities to a regional event: the Iberoamerican School of Astrobiology. His decision granted us the privilege to be involved with a single activity entirely devoted to studying astrobiology in its complete range (Caracas, Venezuela, November-December 1999, cf., Fig. P8).

On that occasion I was able to go deeper in my understanding of astrobiology due to the presence of Drake and Oró, as well as scientists of the region that had already participated in the Trieste conferences; but I also benefitted from the insights of new contacts with other scientists and humanists: Elinor Medina de Callarotti, Humberto Campins, Jesus Alberto Leon, Ernesto Palacios-Pru, Hector Rago, Juan G. Roederer, Sabatino Sofia, Ernesto Mayz Vallenilla and Raimundo Villegas. Altogether 125 participants attended the Caracas meeting (cf., Fig. P9).

The first part of the present text is based, with extensive corrections and modifications, on the contributions made for the 1996 Varenna [2] and the Vatican Symposia[3]. Chapters 1-4 are based on the contributions to those meetings. Chapter 13 is partially based on the contribution to the 1998 Varenna Symposium [4]. Chapters 10 and 11 are partially based on the script of the lectures prepared for the 6th International Conference on Bioastronomy [5], Hawaii, August 1999, as well as the Iberoamerican School of Astrobiology. Chapter 12 is based on an invited lecture at the College on the Evolution of Intelligent Behavior, Trieste, April, 2001.

Figure P8: Closing Ceremony of the Iberoamerican School of Astrobiology. (1999, courtesy of Public Information Office, IDEA).

The inclusion of some intercultural subjects in the present book was suggested to the author while preparing for the lecture delivered at the "Cattedra dei non credenti" (Chair of non-believers) [6]. Chapter 13 is strongly influenced by my participation in that singular and most stimulating event. I would like to thank Roberto Colombo, George Coyne SJ, Mario Gargantini, Giulio Giorello, His Eminence Cardinal Carlo Maria Martini, Robert John which helped me to reflect on deeper aspects of astrobiology. I would also like to express my gratitude to Paul Davies for his advice on matters related to the first draft of this book. I would further like to thank the co-directors of the Trieste Conferences on Chemical Evolution, Tobias Owen and François Raulin for their support and insights they have given me during the course of our work.

In particular, my gratitude must be expressed to George Coyne, SJ, Tobias Owen and Martino Rizzotti for their generous help in detecting some difficulties in earlier versions of the manuscript. Barbara Clancy and Joseph Seckbach also contributed with their timely advice. They also have suggested many ways of improving the presentation, which I have gladly and eagerly incorporated into the present book. But in the extensive process of rewriting, none of the reviewers are responsible for any difficulties that may still remain.

Figure P9: Some of the participants of the Iberoamerican School of Astrobiology (1999, courtesy of Public Information Office, IDEA).

Finally, I would like to thank Sarah Catherine, my wife, for her patience while I was writing this book, for her perennial support and wise counseling.

Julian Chela-Flores
April, 2001
Trieste, Italy.

THE NEW SCIENCE OF ASTROBIOLOGY:

From genesis of the living cell to
evolution of intelligent behavior in the universe

BOOK 1:

ORIGIN OF LIFE IN THE UNIVERSE

Introduction:

What is Astrobiology?

Does life exist beyond our planet and, if so, is that life comparable with life forms we know here on the Earth? These are some of the most fascinating questions facing science today—particularly astrobiology, the study of the origin, evolution, distribution and destiny of life in the universe.

The three strategies in the search for extraterrestrial life

Three strategies have been devised in the search for extraterrestrial life: the study of the cellular makeup of exotic organisms on Earth; the search for organic matter and living microorganisms beyond Earth; and the use of radio telescopes to detect signals of intelligent behavior in the universe.

The first strategy has focused on understanding how life began on Earth. Research has shown exotic organisms living in inhospitable environments, such as the ocean-floor, Antarctic glacial sheets, and volcanic lava streams—all of which display temperatures and pressures that may have been present during the Earth's formation about 4 billion years ago. Perhaps one of the most unexpected recent discoveries has been that there are whole underground ecosystems, which to a large extent are independent of sunlight, extending our old concept of what was a habitable zone in a given solar system. Research into our own origins not only broadens our appreciation of the enormous diversity of life here on Earth, but may help us understand the environmental extremes that simple organisms can tolerate. Such extremes may be found on other celestial bodies making it more probable that life can exist elsewhere.

The second strategy for deciding if we are not alone in the universe is a search for the simplest forms of organic matter—amino acids or proteins—that may be embedded in the ancient rock of planets, comets or meteorites, or in interstellar clouds. The search has focused on three celestial bodies besides our own: Mars, Europa (a moon of Jupiter), and Titan (a moon of Saturn). The discovery of meteorites from Mars suggests that all the terrestrial planets (Mercury, Venus, the Earth, the Moon and Mars) at one stage in the past may have exchanged such material. There is compelling evidence that liquid water has flowed in the geologically recent past on Mars, (or may even be flowing now). The Galileo Space Mission has provided ample evidence for an ocean on Europa underneath its frozen surface. In a few years from now we expect substantial information to unveil the secrets of the hidden surface of Titan, a world with an atmosphere which resembles our own planet when life first emerged on it. The main

5

interest of research on Earth has shifted toward discovering fossilized remnants of life that existed billions of years ago when our world was an inhospitable environment where only thermophilic organisms (those that could withstand extreme temperatures) could survive. One of the primary goals of astrobiology is to determine whether life ever existed in places other than the Earth and, if so, what were the environmental conditions that made it possible.

Research, continues for clues that could reveal the existence of life forms beyond Earth. The discovery of an independent life form, a separate tree of life from our own on Mars or Europa, would not only be fascinating in its own right, but it would shed revealing light on the microorganisms that inhabited the Earth about four billion years ago, when the continual bombardment of meteorites and comets made the surface of the Earth a more hostile environment.

The third strategy used in the hunt for life beyond Earth relies on radio telescopes such as the huge one at the National Astronomy and Ionosphere Center in Arecibo, USA. These huge 'dishes' actually have two roles to play: First and foremost, they help to examine wavelengths that cannot be seen by the human eye; for example, radio waves and microwaves. Such information has proven essential for understanding the movement and behavior of planets and stars. Second, radio telescopes also seek anomalies in microwaves and radio waves. Such anomalies may represent the imprint of intelligent behavior in the cosmos. Thus far astronomers have been scanning the radio spectrum for four decades with no reliable signal from an extraterrestrial civilization. But the absence of positive results does not mean the initiative is likely to be abandoned. The fascination of the search for extraterrestrial life, combined with the vast reaches of outer space, where earth-bound scientific research continues to find new extrasolar planets and even solar systems, keeps the hope alive that 'somewhere out there' there is intelligent life. If a signal is ever received, this much is certain: It will be one of the most remarkable and influential discoveries ever made, whose implications will range from science, philosophy and even theology.

These three strategies have provided a roadmap for a great journey of discovery. As sketched in the following chapters, such an effort is possible due to research currently taking place for finding out the real place of planet Earth in the universe and, at the same time, the place of all life that ever evolved on our small planet (cf., Fig. I.1).

Putting biology back into the space sciences

The present work presents some reflections on astrobiology, in particular on the transition from inert matter to intelligence, as we know it on Earth. It also discusses the possibility of it occurring elsewhere as well, particularly in other environments of solar systems. This possibility may be probed with the forthcoming series of space missions, which at the time of writing are being planned as far as the year 2010 and beyond. Since research in astrobiology is strongly based on the basic sciences, we would like, before beginning the main body of the work, to consider the relation between the basic sciences that serve to support the sciences of the origin, evolution, distribution and destiny of life in the universe.

For simplicity, and in agreement with a growing tendency in the specialized community of researchers, we shall refer to this subject as 'astrobiology', reserving the expressions 'exobiology' and 'bioastronomy' for the specific topics discussed in "Book 3" (cf., Table of Contents, pp. vii - xvi).

Figure I.1:The Earth as seen by the astronauts of Apollo 17
(courtesy of NASA).

In his very influential book [1] Ernst Mayr argues, correctly in our view, against
biology being an imperfect stepchild of the physical sciences. The history of the
classification of living or extinct organisms into groups, a discipline known as
taxonomy is a good point in question, for its traditional functions have been the
identification of particular specimens and the classification of living things into a
satisfactory conceptual order.

In Aristotle's time the basis of such order was not understood. For deep
insights we had to wait for the advent of the birth of modern biology in the middle of
the 19th century, with the work of Charles Darwin. In *The Origin of Species* [2] a true

revolution was brought about by incorporating a very wide range of biological phenomena within the scope of natural explanation, including an appropriate approach to the study of 'taxonomy', namely, the theory, practice and rules of classification of living or extinct organisms into groups, according to a given set of relationships.

Darwin's theorizing in biology may have begun in 1838 (in his Notebook D), although for reasons that have been well documented elsewhere, the publication of his seminal book was delayed until 1859.

The intellectual revolution initiated by Darwin has produced a far-reaching change in our thinking, even outside the strict boundaries of the natural sciences, but the radical departure from the biological science of his time consisted in the formulation simultaneously with Alfred Wallace of a mechanism, natural selection, for the evolution of life.

At a time when the prevalent ideas in the scientific community were mathematical principles and physical laws, biology was enriched with the concepts of chance and probability. In spite of the fact that it is widely appreciated that biology is not the imperfect stepchild of the physical sciences, physics itself nevertheless has had a valuable influence on the life sciences, particularly when a century after the publication of *The Origin of Species,* the molecular bases of genetics were searched for with precise experiments.

A   role for physics in the birth of modern biology

During the 1940s physicists and physical chemists made vital contributions to the transition of biology into an experimental phase, in which traditional physical sub disciplines were to make modern molecular biology possible. The theoretical physicist, Max Delbrück, had contacts with radiation biologists [3]. Learning about mutations, he foresaw the possibility of linking them to the lesions produced by a physical agent acting on the atoms of the still-to-be-discovered genetic material. This important step in the development of modern science was popularized by the theoretical physicist, Erwin Schrödinger, working during that decade at the Dublin Institute for Advanced Studies. He delivered at Trinity College, Dublin, a set of lectures, known to us through the book *What is Life?* [4]. In those early years, when science was still ignorant of the chemical nature of the carrier of genetic information, a generation of physicists was responsible for the birth of the new biology, together with colleagues trained in the more traditional branches of the biological sciences. This group included, Sir Francis Crick, who graduated in physics at University College, London, whom Lewis Wolpert has referred to as "the genius theoretical biologist of our age" [5].

The origin of life is currently one of the most attractive interdisciplinary fields. In spite of dealing with the origin of biology, major components come from the physical, as well as the earth sciences. It is not difficult to understand the reasons for the ever-growing popularity of the field. They are implicit in the facts that are gradually emerging regarding the first steps of life on Earth.

The role of the earth sciences in the study of the origin of life

Colonies of bacterial microfossils were discovered by J. William Schopf in Western Australia in the early 1990s. A date of some 3500 million years before the present

(Myr BP) may be assigned to these fossils. The structures are called stromatolites and today they are formed by a group of very ancient microorganisms called cyanobacteria [6].

However, according to geochemical work of Manfred Schidlowski [7] and others since then, including Gustaf Arrhenius and co-workers, it is not to be excluded that photosynthetic bacteria might be as old as some of the oldest extant rocks available from the Isua peninsula in Greenland. These are amongst the oldest rocks on Earth; the first identification of these rocks was due to Stephen Moorbath and others. Their work has contributed to a long, still ongoing process of retrieving rocks dating from 3800 Myr BP from a very ancient geologic era, the early Archean. This date should be compared with the age of the Earth itself, of some 4600 Myr BP [8].

According to Schidlowski the 'smoking gun' is the carbon content of the fossils available, since they are biased towards an isotope of carbon, which is normally left behind as the result of the metabolism of cyanobacteria. (Such isotopic enrichment is characteristic of those bacteria that are able to sustain themselves through the process of photosynthesis.) What is clear from the work of the above specialists in the earth sciences is the following: micropaleontologists, biogeochemists and geochronologists have argued that the early Earth biota was dominated by cyanobacteria. As we mentioned above, these are prokaryotic microorganisms, which were at one time called blue-green algae. A transcendental aspect of their physiology is that they were able to extract hydrogen from water in photosynthesis liberating oxygen.

Cyanobacteria were so successful that, as time passed, free oxygen accumulated in the atmosphere, thereby transforming radically the primitive Earth atmosphere, which according to most scholars originally resembled the one we observe today on Titan.

We shall return to this interesting satellite in Chapter 9. Suffice it to say at this stage that during the early geologic history of the Earth, the reactants freely available in the environment, such as iron, had been used up. This showed up in the geologic record as a transition from "banded iron" formation (this will be discussed in more detail in Chapter 5) to rocks largely colored in red. This second group of rocks are called "red beds".To give a compelling name to the colossal transformation of the Earth, this increment in atmospheric oxygen has appropriately been referred to as the 'oxygen holocaust' [9], in order to convey the idea that microorganisms, other than cyanobacteria, had no defenses against the new atmosphere that was developing. The only passive remedy for life on Earth was evolution into new forms that tolerated, and even thrived on oxygen. This led to the onset of the highly evolved cellular blueprint (eukaryotic), which will be the main theme of this book.

Organized studies and meetings on the origin of life

The identification of the mechanism for the initiation of life has been in the minds of philosophers and theologians for as long as civilization has existed. In modern times, with the pioneering work of Alexander Oparin, the discipline of the study of the origin of life has reached maturity as a research field, and the origin, evolution and distribution of life in the universe has become a valid topic of interdisciplinary research.

As we have seen in the Preface, the personal interest of Abdus Salam, in the wide area of theoretical biology was a key factor in the introduction of the conferences on chemical evolution in Trieste. He gave a strong impulse to these studies with his own research in the topic of the symmetry of the macromolecules of life, discussed in his publications from 1990 till 1993 [10].

The first unifying principle in biochemistry is that the key molecules: amino acids, sugars, and phospholipids, have the same 'handedness' (a property of molecules that exist in two forms, whose spatial configurations are mirror images of each other. They are referred to as left (L) or right (R) configurations. The particular biomolecules that are most often discussed from this point of view are the amino acids, but sugars and phospholipids show analogous properties. Remarkably, this is true for all organisms with the exception of bacterial cell walls, which contain D-amino acids, as in the case of *Lactobacillus arabinosus* [11]. However, we may state in general that living systems translate their genes into proteins composed of twenty L-amino acids, following certain rules (cf., Chapter 4 "the genetic code"). Since the 1950s the origin of molecular handedness has been searched for in the effects of the weak subnuclear force amongst elementary particles [12] (now called 'electroweak interaction'). In a substantial body of previous work by many authors, a component of the electroweak subnuclear force has been suggested as the main physical factor inducing observed molecular handedness.

The subject of the biological handedness of the macromolecules of life (chirality) was the focus of attention of Salam's last research topic. His approach also invoked the electroweak interaction, but was original in appealing to further physical concepts that may apply at the end of chemical evolution. With physical concepts Salam was led to suggest that the electromagnetic force is not the only force that can produce chemical effects: he argued that a component of the electroweak force, in spite of the fact that its effects appear to be negligible at low energies, may play an active role in chemistry.

The reasons for the proposed chemical role of the parity-violating weak interactions [13] may be found in some calculations in quantum chemistry [14] of Stephen Mason, Alexandra MacDermott and George Tranter working in the United Kingdom. Independent calculations of Ayaz Bakasov, Tae-Kyu Ha and Martin Quack from Zürich gave some support to the previous assumptions that Salam had used from quantum chemistry in his own work. Cyril Ponnamperuma, who was also attracted to this original approach to the question of chirality, proceeded to test these ideas in his own world-renowned Laboratory of Chemical Evolution at the University of Maryland. Although these experiments failed to confirm the model, the experiments were continued by other groups, particularly by Wang Wenqing and co-workers in Beijing.

Today, the robust growth of the field includes very accurate verification that organic contents of some ancient meteorites show the same asymmetry as the molecules of the living cell (cf., the Murchison meteorite, Chapter 3.) These results were discovered in the later part of the 1990s. They suggest a scenario in which organic matter of extraterrestrial origin has played an essential role in the mystery of why the protein amino acids of the living world are left-handed.

The ideas of the early 1990s, which attempted to pin down the source of biological handedness to one of the subnuclear interactions now find their natural context in a cosmic setting, rather than on the surface of the primitive Earth. In the following chapters we shall benefit form the extensive reviews of the subject included in the Trieste conferences [15-20], and the Iberoamerican School of Astrobiology [21].

Part I

CHEMICAL EVOLUTION: FOUNDATIONS FOR THE STUDY OF THE ORIGIN OF LIFE IN THE UNIVERSE

1

From cosmic to chemical evolution

Theologians were raising questions that we still cannot answer fully, long before the advent of science and philosophy. The first steps of philosophy were taken a few centuries before our own era by the ancient Greeks, notably by Thales of Miletus (active in 585 BC). A significant contribution to the birth of science was made by Democritus of Abdera, one of the founders of atomism (active in 420 BC).

It is important to realize that the Bible itself is even more ancient as, for instance, the Prophet Jeremiad, a contemporary of Thales, was active during the destruction of Jerusalem in 587 BC, which had been founded by King David almost half a millennium earlier. Besides, Ezra, who gave Judaism its distinctive character by his promulgation of the by then ancient Mosaic Law in 430 BC, was a contemporary of Democritus.

Some cultural questions

Some of the deepest questions that have persistently remained with us since biblical times are:

What is the origin of the universe?
What is it made of?
What is its ultimate destiny?
How did life, in general, and humans, in particular,
*   originate?*
Are we alone in the cosmos?

13

I feel that there is no better way to organize the information available to scientists, than to return to those questions as originally set in the antiquity, in order to appreciate, in an orderly manner, the considerable progress that has been achieved up to the present in our understanding of cosmological models and the appearance of intelligent life on Earth. One possible way of retrieving these eternal questions is by entering the atrium of St. Mark's Basilica in Venice. As you go through St. Clement's Portal in its South-Western corner, you may observe a small cupola, illustrating the initial events of the Book of Genesis (cf., Fig. 1.1).

Fig. 1.1: The cupola in the West Atrium of St. Mark's Basilica in Venice. (With kind permission of the Procuratoria of St. Mark's Basilica.)

This is only part of a complete 13th century overview of the Old Testament, which continues with a cycle of five other cupolas along the western and northern sides of the Basilica. The Genesis cupola itself is divided into 24 scenes set on three concentric bands. The iconography follows an Alexandrian illuminated manuscript miniated during the 5th century, which is known as the "Cotton Bible", as parts of it were donated in the 18th century to the British Museum by Sir John Cotton. (This codex now belongs to the British Library in London, since its foundation in 1973.)

The innermost band of the cupola, and part of the middle band, raise some of the questions that science has attempted to answer. We will address this problem in the first three sections of this work. In Chapter 6 we pick up the story of the origin of multicellular life and the onset of the evolutionary process that led to the appearance of intelligent behavior on Earth. The middle and outer bands of the Genesis cupola already touch on this event by raising the questions of the first appearance of fish, birds and, in separate scenes, mammals and humans. The topics we discuss in this book go part of the way over the range of questions raised in this complete cycle of six cupolas.

However, in spite of the impressive progress of science, it is important to underline that many of the fundamental questions in astrobiology still remain unanswered. Therefore, before beginning our task it may be prudent to recall that cosmology suggests that there is sufficient time available for science to progress to a stage in which further, and possibly better attempts than are now possible, will be made, in order to search for answers to some of the deepest questions (mentioned above) which humans have been asking themselves from time immemorial. Up to the present partial answers have often led to controversy between scientists, philosophers and theologians. We should be made aware of the limited scope of the scientific method, a point that has been stressed by Bertrand Russell. Recognizing a limit to the applicability of present day science, he was of the opinion that almost all the questions which interest speculative minds are of such a nature that science is at present unable to answer [1].

Cosmological models

Our first topic is suggested by the Book of Genesis ("*In the beginning God created the heavens and the earth*" : **1**, 1). To approach this problem from the point of view of science, a preliminary step is to grasp the significance of the scale of time involved. For this purpose we must return to the first instants of cosmic expansion [2].

The American scientist Edwin Hubble discovered in 1929 that large groups of stars (galaxies), which in addition contain interstellar matter and nebulae, were moving away from us and that the velocity V of recession of such galaxies is proportional to their distance d from us. (This is known as Hubble's Law.) This implies if we assume isotropy that the universe as a whole is expanding. The distances that measure this phenomenon are normally expressed in megaparsecs (1 pc $=$ 3.26 ly; 1 ly, in turn, denotes the distance that light travels in a year; 1 megaparsec, Mpc $= 10^6$ pc). In other words, $V = H_O d$. The constant of proportionality, which is known as the Hubble constant H_O is, consequently, given by the ratio of the speed of recession of the galaxy to its distance; H_0 represents quantitatively the current rate of expansion of the universe. Determination of the value of H_0 has been difficult and the true value is still the subject of controversy [3]. A growing consensus puts H_0 in the range 60-80 km s^{-1} Mp^1.

The current estimates of the Hubble constant in the standard cosmological model (cf., the Friedmann model below) of an expanding universe imply an age of the universe in a range of over 10 thousand million years (Gyr), the precise values depending on the particular assumption we may adopt for the matter density present in the universe. An alternative approach measures the abundance of radioactive elements found in stars. Astronomers search for very old stars and estimate the abundance of a given radioactive element. From nuclear physics the time of decay of the element allows an estimate of the age of the galaxy itself. This method uses in particular the elements uranium and thorium. The current estimate puts the age of the universe at around 12.5 Gyr) [4]. Unfortunately, this state of affairs raises some difficulties, as there are globular clusters in our galaxy whose age is estimated to be 13-17 Gyr. Ours is a typical spiral galaxy, not unlike others that are well known form the images provided by the Hubble Space Telescope (cf., Fig. 1.2).

On the other hand, life on Earth extends back only to some 4 Gyr before the present (BP), while the age of the solar system is 4.6 Gyr BP. From the point of view of the origin of life on Earth we therefore do not need to follow the cosmological arguments in detail. The chemical evolution scenario faces no particular difficulties with the above values of H_0

Fig. 1.2: The large spiral galaxy NGC 4414 image completed in 1999 (courtesy of NASA/HST).

The theory of gravitation formulated by Albert Einstein is known as General Relativity (GR). Cosmological models may be discussed within the context of this theory in terms of a single function R of time t. This function may be referred to, quite appropriately, as a 'scale factor', a measure of the size of the universe.

Sometimes, when referring to the particular solution the expression 'radius of the universe' may be preferred for the function R. As the universal expansion sets in, R is found to increase in a model that assumes homogeneity in the distribution of matter (the 'substratum'), as well as isotropy of space. The functional dependence of R, as a function of time t, is a smooth increasing function for a specific choice of two free parameters, which have a deep meaning in the GR theory of gravitation, namely, the curvature of space and the cosmological constant. The functional behavior of the scale factor R was found by the Russian mathematician Alexander Friedmann in 1922. This solution is also attributed to Howard Percy Robertson and A.G. Walker for their work done in the 1930s. Such a (standard) model is referred to as the Friedmann model. In fact, R is inversely proportional to the substratum temperature T.

Hence, in the context of the expansion of the universe we have been discussing, since R is also found to increase with time t (cf., the previous paragraph), T decreases; this model implies, therefore, that as t tends to zero (the 'zero' of time) the value of the temperature T is large. (The temperature goes to infinity as t tends to zero.) In other words, the Friedmann solution suggests that there was a 'hot' initial condition. As the function R represents a scale of the universe (in the sense we have just explained), the expression 'big bang', due to Sir Fred Hoyle, has been adopted for the beginning of the universe in the Friedmann model. The almost universal acceptance of big bang cosmology is due to its observational support.

The cosmic microwave background

The big bang model tells us that as time t increases, the universe cools down to a certain temperature, which at present is close to 3° K. This discovery took place in 1964 by Arno Penzias and Robert Wilson; they provided solid evidence that the part of the universe surrounding us is presently illuminated by "$T = 3^o$ K" radiation, the cosmic background, but since it has a typical wavelength of about 2 mm, it is referred to as the cosmic 'microwave' background, CMB. It may be confidently considered to be a cooled remnant from the hot early phases of the universe.

It has an 'isotropic' distribution, in other words, its temperature does not vary appreciably independent of the direction in which we are observing the celestial sphere (the accuracy of this statement is 10 parts per million, ppm). The isotropy is a consequence firstly of the uniformity of cosmic expansion, secondly, of its homogeneity when its age was 300,000 years and temperature of 3,000K [4].

On the other hand, in 1992 more precise measurements of the $T = 3^o$ K" radiation, began to be made by means of the satellite called the Cosmic Background Explorer (COBE):

When the accuracy of the isotropy was tested with more refined measurements, it was found that there was some degree of anisotropy after all - the temperature did vary according to the direction of observation (one part in 100,000). This fact is interpreted as evidence of variations in the primordial plasma, a first step in the evolution of galaxies.

Further accuracy in understanding the deviations from isotropy of the CMB (and hence a better understanding of the early universe) can be expected in the next few years. The Microwave Anisotropy Probe (MAP) - an initiative of the National Aeronautics and Space Administration (NASA) - should extend the precise observations of the CMB to the entire sky. MAP will be in a solar orbit of some 1,5 million kilometers. This will be extended subsequently by the European Space Agency (ESA) by means of the Planck spacecraft, planned for the year 2007.

Further views on the cosmos

Andre Linde envisions a vast cosmos in a model in which the current universal expansion is a bubble in an infinitely old 'superuniverse'. In other words, in this model the universe we know is assumed to be a bubble amongst bubbles, which are eternally appearing and breeding new universes. Final observational confirmation is still missing.

The scale parameter R evolves as a function of time t in such a cosmology, but as in the earlier model of Alan Guth, the Linde model differs from the Friedmann solution in the first instants of cosmic evolution. Depending on the way the word "standard" is used, both the Linde and Guth models can be considered standard big bang models.

The unexpected repulsive cosmic force

We should conclude this brief overview of cosmology with a note of caution. The theoretical framework described here may have to be revised in order to take into account careful measurements of the velocities of very distant galaxies as defined by their stars in their late stage of evolution, namely, exploding stars that have exhausted their nuclear fuel. These important dying stars are normally called supernovae [5]. The value of these velocities may be interpreted as some evidence for an accelerating expansion of the universe, a phenomenon, which is still to be understood. We have to learn whether the constant that Einstein introduced into his equations of gravitation (the 'cosmological constant'), purely on theoretical grounds, may represent some form of gravitational repulsion, rather than attraction [6].

We should dwell on this question a little longer. Only a small fraction of the matter in the universe is in the form of the familiar chemical elements found in the Periodic Table. It is assumed that a large proportion of the cosmic matter consists of 'dark matter', whose composition consists of particles that play a role in the sub-nuclear interactions, mostly foreign to our everyday experience. The term 'dark matter' is not a misnomer, for the sub-nuclear particles that contribute to it, do not interact with light.

However, a remarkable aspect of cosmic matter is emerging: the sum total of the standard chemical elements and the dark matter make up only half of the matter content of the universe. The remaining fraction of cosmic matter has been referred to as dark energy with the astonishing property that its gravity is repulsive, rather than attractive.

A possibility that has to be considered seriously in the future is that the repulsive gravity may dominate the overall evolution of the universe. This could lead to ever increasing rates of expansion. If this was to be the future of our cosmos, then future of life in the universe may hold some surprises. Eschatological considerations may have to be revisited. Will our universe be biofriendly?

But this has taken us far enough into the general picture of the origin and evolution of the universe and within this framework we may begin to consider how life was inserted in the universe in the first place.

The birth of studies on the origin of life

Science today aims at inserting biological evolution in the context of cosmological evolution. We will mainly attempt to convey the idea of an ongoing transformation in studies on the origin of life. This progress in understanding our own origins began in the middle of the 20th century, triggered by the success in retrieving some key biomolecules in experiments which attempted to simulate prebiotic conditions. Some of the main experiments of the 1950s and early 60s were done by organic chemists including Melvin Calvin, Stanley Miller, Sidney Fox, Joan Oró, Cyril Ponnamperuma and their co-workers. Since that time the field has continued its robust growth.

These efforts have led us to view the cosmos as a matrix in which organic matter can be inexorably self-organized by the laws of physics and chemistry into what we recognize as living organisms. However, it should be stressed that chemical evolution experiments that have been carried out to this point have been unable to reproduce the complete pathway from inanimate to living matter. This aspect of chemical evolution has been repeatedly pointed out by Robert Shapiro [7].

Thus, at present the physical and chemical bases of life that we have sketched are persuasive, but further research is still needed. Originally the subject began to take shape as a scientific discipline in the early 1920s [8], when an organic chemist, Alexander Oparin, applied the scientific method of hypotheses and experiments to the origin of the first cell, thereby allowing scientific enquiry to shed valuable new light on a subject that has traditionally been the focus of philosophy and theology.

From Darwin's seminal insight, *The Theory of Common Descent* [9], we may say that "probably all the organic beings which have ever lived on this earth have descended from some primordial form, into which life was first breathed". Darwin's insight offers a unifying explanation for the existence of life on Earth, and allows for meaningful questions regarding early terrestrial evolution. The ancestry of three species of Galapagos mockingbirds that had clearly descended from a single ancestral species on the South American continent, convinced Darwin that all organisms on Earth descended from a common ancestral species.

With the enormous scope of bacterial evolutionary data available today from an extensive micropaleontological record, the *Theory of Common Descent* leads back to a single common ancestor, a progenote or "cenancestor" [10].

The characteristics of the cenancestor can be studied today through comparison of macromolecules, allowing researchers to distinguish and recognize early events that led to the divergent genesis of the highest taxa (domains) among microorganisms. Strictly speaking, in the 'intermediate period', ranging from the first living cell to the cenancestor, life may have evolved in the absence of significant diversity, effectively as a single phylum, incorporating organisms whose genetic systems were already based on DNA [11]. Earlier still, prior to the encapsulation of nucleic acids in microspheres, evolution may already have been at work on RNA molecules (the 'RNA world').

Chapters 2 to 4 are devoted to the discussion of certain concepts that may have played a relevant role in the pathway that led to the origin and evolution of the cenancestor. For practical purposes, in the present discussion we use the working

definition of life adopted by the National Aeronautics and Space Agency (NASA) [12]:
Life is a self-sustained chemical system capable of undergoing Darwinian evolution.

The growth of studies on the origin of life

The still ongoing second large step in the development of origin of life studies consists
of taking the subject from the laboratories of the 'card-carrying' organic chemists, where
Oparin had introduced it, into the domain of the life, earth and space scientists. In fact,
the subject of astrobiology has come of age due to the space missions of the last 30
years: Mariner, Apollo, Viking, Voyagers and particularly the most recent ones,
Galileo, which is currently sending valuable information on the Jovian system and
Cassini-Huygens that will soon yield its first results (cf., Table 1.1). Further missions
are being planned.

TABLE 1.1 : Some of the present and future space missions [13].

Mission	Launch	Cruise end	Mission end
Mars Global Surveyor	7/11/96	9/5/97	31/1/00
Mars Pathfinder	2/12/96	4/7/97	4/8/98
Cassini-Huygens Mission	15/10/97	1/7/04	1/7/08
Mars Global Surveyor-Climate Orbiter	10/12/98	23/9/99	Mission lost
Mars Global Surveyor-Polar Lander	8/1/99	3/12/99	Mission lost
Stardust (comet dust sample return)	6/2/98	15/1/06	15/1/06
Rosetta (comet lander)	1/7/03	1/8/11	1/7/13

New areas of research have come into existence as a consequence of this change
of emphasis, such as *planetary protection,* whose aim is to prepare ourselves in advance
for the sample-return missions from Mars and Europa; in the case of Mars, this event

cannot take place as early as the year 2005, as originally intended. It should be stressed that there are two kinds of planetary protection:
• Protecting other planets from us. This has been a requirement in previous missions such as the Viking mission to Mars in 1976.
• Protecting us from possible exogenous contamination. This is needed only in sample-return missions, as discussed here.

Experiments and sample retrieval missions from Europa are very relevant. Indeed, an early NASA and National Science Foundation (NSF) meeting was convened in 1996 to discuss such matters (*Europa Ocean Conference*, San Juan Capistrano, California). Since then other specialized meetings have maintained their focus on Europa. The subject of planetary Protection has been pioneered by Donald DeVincenzi at NASA-Ames, John Rummel at the Marine Biological Laboratory, and Margaret Race at the SETI Institute. A second discipline that has arisen from the exploration of the Solar System is *comparative planetology* [14], a new subject which allows us to learn more about phenomena in each specific planet, or satellite, by having several examples to examine. A good illustration is the phenomenon of volcanism, which is remarkably different in the Jovian satellite Io, as compared to what is observed on Earth. These studies are needed in the formulation and development of devices, which may allow the search for Solar System relics of the earliest stages of the evolution of life.

The problem of the search for extraterrestrial life may be approached in order of increasing complexity:
a) At the organic and inorganic level the search for extraterrestrial homochirality has been advocated. Homochirality refers to single-handedness observed in the main biomolecules, namely, amino acids, nucleic acids and phospholipids. Consequently, in an extraterrestrial environment testing such properties will be of considerable interest, since molecular chirality may be considered as an important biochemical biomarker.
b) The search of microfossils of bacteria in Martian meteorites has been another alternative, which we shall consider later on (cf., Chapter 3).
c) In Chapter 8 we shall argue that the search for extraterrestrial single-cell organisms of a nucleated type is an important possibility in future missions. (A word of Greek origin 'eukaryote' is normally assigned to such truly nucleated cells, cf., Chapter 5.)
d) Finally, the search for extraterrestrial intelligence (SETI) was pioneered by Frank Drake [15]. The eventual success of SETI may depend on the adoption of truly significant enterprises, such as its extension to the far side of the Moon (in the Saha crater), as advocated by the late Jean Heidmann from the Observatory of Paris, Claudio Maccone from the G. Colombo Centre for Astrodynamics in Turin and Giancarlo Genta, also from Turin (Politecnico di Torino). In Chapter 11 we shall return to SETI and, in particular, to "SETI on the Moon". In Chapter 12 we shall discuss some theoretical bases for the likelihood of the evolution of intelligent behavior elsewhere; we shall also discuss experimental means for testing the first steps in such evolutionary pathways. The construction of a road linking laboratories in Mare Smythii (at the equatorial level and at the Moon's limb) and the Saha crater has been proposed [16].

None of these engineering projects seem beyond present technological capabilities. In fact, the only limitation foreseen for obtaining further insights into the nature of the origin of life is the will of our society to maintain the present rate of progress, by providing adequate budgets to scientists and engineers, who are involved in the main programs of astrobiology.

To continue our overview of the origin of life in the universe, we must first return to cosmological models.

Origin of the elements: from the big bang to the interior of stars

As we have seen above, according to the big bang model, initially the temperature of the primordial matter was so elevated that atomic, nuclear or subnuclear stability was impossible. In less than one million years after the beginning of the general expansion, the temperature T was already sufficiently low for electrons and protons to be able to form hydrogen atoms. At an earlier epoch there would have been too much thermal energy for the electromagnetic force to be able to bound up nuclei and electrons.

Up to that moment these elementary particles were too energetic to allow atoms to be formed. Once 'recombination' of electrons and protons was possible, due to falling temperatures, thermal motion was no longer able to prevent the electromagnetic interaction from forming hydrogen atoms. This is the 'moment of decoupling' of matter and radiation. At this stage of universal expansion the force of gravity was able to induce the hydrogen gas to coalesce into stars and galaxies.

A series of nuclear reactions in the interior of stars was proposed by Hans Bethe. His aim was to understand the nuclear reactions that are responsible for the luminosity of the Sun and other stars. [17].

The subsequent development of this theory explains how all the elements we know can be built up through nuclear processes inside stars, and by giant stellar explosions. The underlying phenomenon consists of high energy collisions between atomic nuclei and elementary particles that have been stripped off their corresponding atoms, or even nuclei, due to the presence of the enormous thermal energy in the core of the star.

At such high temperatures nuclear fusion may occur; in other words, there can occur nuclear reactions between light atomic nuclei with the release of energy. In the interior of stars reactions are called thermonuclear when they involve nuclear fusion, in which the reacting bodies have sufficient energy to initiate and sustain the process.

One example is provided by a series of nuclear reactions that induce hydrogen nuclei (essentially single 'elementary' particles called protons) to fuse into helium nuclei. A helium nucleus is heavier than the proton; in fact, it consists of two protons and two slightly more massive particles called neutrons. But the total mass of the helium nucleus is actually a tiny bit less than the sum of the masses of the four particles. In the process of forging the nucleus from its components, the missing mass is converted directly into energy, following Einstein's famous equation, $E = mc^2$, where E denotes energy, c is the velocity of light and m is the mass. This is the energy that produces sunlight.

This process, in addition, releases other particles and some energy. After a long phase (measured in millions of years) dominated by such conversion, or 'burning' of hydrogen into helium, the star evolves: its structure becomes gradually that of a small core, where helium and the heavier elements accumulate. Since both temperature and density of the core increase, the pressure balance with the gravitational force is maintained.

The star itself increases in size. It becomes what is normally called a "red giant", because at that stage of their evolution they are changed into a state of high luminosity and red color. During this long process, from a young star to a red giant, the elements carbon and oxygen, and several others, are formed by fusing helium atoms. (cf., Table 1.2).

Supernovae: the source of biogenic dust

As we shall see later, solar systems originate out of interstellar dust, which can be considered mostly as 'biogenic dust', namely dust constituted mainly out of the fundamental elements of life such as carbon (C), hydrogen (H), oxygen (O) and nitrogen (N), and a few others.

It is for this reason that the late stage of stellar evolution is so relevant. In fact, just before a star explodes into its supernova stage, all the elements that have originated in its interior out of thermonuclear reactions are expelled, thus contributing to the interstellar dust. At a later stage in the stellar evolution the expanding dust and gas forms typical 'planetary nebulae (cf., Fig. 1.3).

Let us consider next some of the details of these 'life-generating' astrophysical events. Indeed, stars whose mass is similar to that of the Sun remain at the red giant stage for a few hundred million years. The last stages of burning produce an interesting anomaly: the star pushes off its outer layers forming a large shell of gas; in fact, much larger than the star itself. This structure is called a planetary nebula. The star itself collapses under its own gravity compressing its matter to a degenerate state, in which the laws of microscopic physics (called quantum mechanics), eventually stabilize the collapse. This is the stage of stellar evolution called a 'white dwarf '. The stellar evolution of stars more massive than the Sun is far more interesting from the point of view of astrobiology. After the massive star has burnt out its nuclear fuel (in the previous process of nucleosynthesis of most of the lighter elements listed in Table 1.2), a catastrophic explosion follows in which an enormous amount of energy and matter is released.

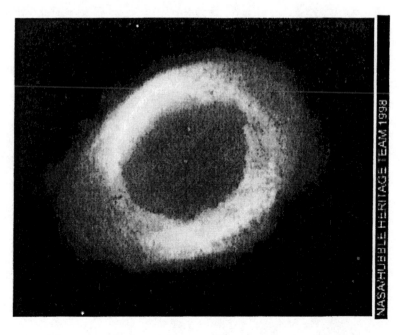

Figure 1.3: The Ring Nebula (M57). The figure shows gas and dust that has been cast off thousand of years after the death of the star (courtesy of NASA).

It is precisely these 'supernovae' explosions that are the source of enrichment of the chemical composition of the interstellar medium. This chemical phenomenon, in turn, provides new raw material for subsequent generations of star formation, which in some cases leads to the production of planets.

Late in stellar evolution stars are still poor in some of the heavier biogenic elements (such as, for instance, magnesium and phosphorus). Such elements are the product of nucleosynthesis triggered in the extreme physical conditions that occur in the supernova event itself. By this means the newly synthesized elements are disseminated into interstellar space, becoming dust particles after a few generations of star births and deaths (cf., Chapter 2, Cosmochemistry).

Molecular clouds and circumstellar disks

As a result of several generations of stellar evolution, a large fraction of the gas within our galaxy is found in the form of clouds that originally consisted only of molecular hydrogen, but now include many of the heavier elements.

The liner dimensions of these clouds can be as large as several hundred light years. The masses involved may be something in the range 10^5 to 10^6 solar masses.

TABLE 1.2 : Chemical composition of the Sun (adapted from ref. 18).

Element	Symbol	Percentage
Hydrogen	H	87.000
Helium	He	12.900
Oxygen	O	0.025
Nitrogen	N	0.020
Neon	Ne	0.030
Carbon	C	0.01
Magnesium	Mg	0.003
Silicon	Si	0.002
Iron	Fe	0.001
Sulfur	S	0.001
Others		0.038

It is useful to consider one example in detail. A large molecular cloud in the constellation of Orion has a few hundred thousand solar masses; its diameter is about 1500 light years [19]. Its temperature ranges from 10 to 50 K. Within the last two million years a few stars have been born there. The most recent observations demonstrate that star formation continues there even today.

Images from the Earth-orbiting Hubble Space Telescope (HST) have shown circumstellar disks surrounding young stars. This supports the old theory of planetary formation from a primeval nebula surrounding the nascent star. In the case of our solar system this gas formation has been called the 'solar nebula'. As we shall see in the next chapter, the coplanar orbits of the planets together with the direction of their angular momentum [20], coinciding with the direction of the solar angular momentum, argue in favor of the hypothesis of the solar nebula.

The HST images indicate, in exact agreement with the hypothesis, the presence of opaque disks of gas and dust. What is more important, the estimated sizes of the observed circumstellar disks are compatible with the size of the Solar System.

The abundance of elements in the solar nebula is an important issue that will be relevant later on. In general we may say that solar-nebula abundance should coincide to a large extent with solar values (cf., Table 1.2). There are some small differences that will be due to the nuclear-reaction processes taking place in the Sun.

In addition, there will be some differences in the abundance observed in the planets, because as the nebula evolves into a solar system the elements themselves will be partitioned according to simple physicochemical laws. In strict thermodynamic equilibrium carbon would have reacted with oxygen producing exclusively carbon monoxide (CO) in the inner solar system due to the higher thermal energy available in this region [21]. The cooler parts of the outer solar nebula would have favored the formation of methane. However, the testimony of more complex forms of carbon present in comets and meteorites suggests that the parent molecular cloud had already had a condition favorable for chemical reactions that lead to the formation of hydrocarbons [22]. Indeed, observations at radio wavelengths that can penetrate those dark molecular clouds have revealed about 100 different kinds of molecules that are formed there. This sketch of the pathway to carbon compounds and its partition during planetary formation will be very significant in the following chapters (particularly in Chapter 8): it will help us to trace the abundance of organic compounds in sites of the Solar System where life could arise.

It is worthwhile to point out at this stage that in 1983 the Infrared Astronomical Satellite (IRAS) discovered some stars with excess thermal radiation not accountable by the amount that the star could emit. The result was interpreted in terms of disks of dust around these stars. The most famous example is the second brightest star of the southern celestial hemisphere constellation of "The Painter's Easel" (Pictor). According to tradition, the star is called by the second letter of the Greek alphabet and the name in Latin: "Beta-Pictoris" (β-Pictoris, or β-Pic, at a distance of 59 light years).

More recently, a highly unusual example of one of these circumstellar disks has been obtained with an infrared camera (the Near Infrared Camera and Multi-Object Spectrometer (NICMOS) mounted on the HST [23]. The disk is a ringlike structure with characteristic width of 17 astronomical units (the distance equivalent to the radius of the Earth's orbit: it is abbreviated as AU); the inner and outer radii are respectively about 55 AU and 110 AU.

This suggests that the constraint that distorts the disk might be the presence of planets. This is remarkable since the star (catalogue name HR 4796A) at the center of the ring is young (8 million years, Myr), at a distance some three and a half times further away than the star β-Pic. (The precise position had been determined by HIPPARCOS, the European astrometric satellite, launched in 1989 and used for 4 years.)

Both HR 4796A and β-Pictoris are blue-white stars (technically called type A) like Vega, the brightest star in the constellation Lyra, well known in the night skies of the Northern hemisphere. The dust distribution has been found to be analogous to the Kuiper Belt and zodiacal components of dust within our solar system [23].

Stellar evolution

We have already seen that stars evolve as nuclear reactions convert mass to energy. In fact, stars such as our Sun follow a well known pathway along a *Hertzprung-Russell* (HR) diagram (cf., Fig. 1.4). This diagram was introduced independently by Ejnar Hertzprung and Henry Norris Russell in 1913. They observed many nearby stars and found that the plot of their luminosity (i.e., the total energy of visual light radiated by the star per second), and surface temperatures, the plot exhibits a certain regularity: Indeed, the stars lie on the same curve in a diagram (the HR diagram), whose axes are the two parameters considered by the above-mentioned early 20th-century astronomers. Such stars are called *main sequence* stars. It is interesting to remark that the surface temperature of the star is indicated by its "Planckian" radiation [24], which is measured by its spectral type, or color index. We may ask:

How do stars move on the HR diagram as hydrogen is burnt?

Extensive calculations show that main sequence stars are funneled into the upper right hand of the HR diagram, where we find red giants of radii that may be 10 to 100 times the solar radius. Stellar evolution puts a significant constraint on the future of life on Earth, since the radius of the Earth's orbit is small. (Since the eccentricity of the Earth orbit is small, we may speak to a good approximation of a circular orbit, instead of an elliptical one.) On the other hand, the theory of stellar evolution tells us that the radius of the Sun is bound to increase as it evolves off the main sequence in the HR diagram. In the case of the Sun the expected growth of the solar radius will be such that the photosphere will reach the Earth's orbit [25], thus eventually ending life on Earth. The current estimate is that life may continue on Earth for another 4-5 Gyr before the Earth's orbit is no longer in what we may call a 'habitable zone' [26]. However, the implication of the remaining period of habitability of the Earth is profound for our species, as suggested in the next section.

Implications of stellar and biological evolution

According to the standard view of paleoanthropology, the species to which humans belong is referred to as *Homo sapiens sapiens* in order to distinguish us from an independent hominid line which became extinct some 40,000 years ago, i.e., *Homo sapiens neanderthalis* In fact, humans have evolved in less than a few million years since the last common ancestor of the hominids. This common ancestor (technically known as a hominoid) must have lived just some 5 Myr BP [27]. Such a fast evolutionary tempo has occurred within about 1/1000 of the geologic time that is still available for life on Earth.

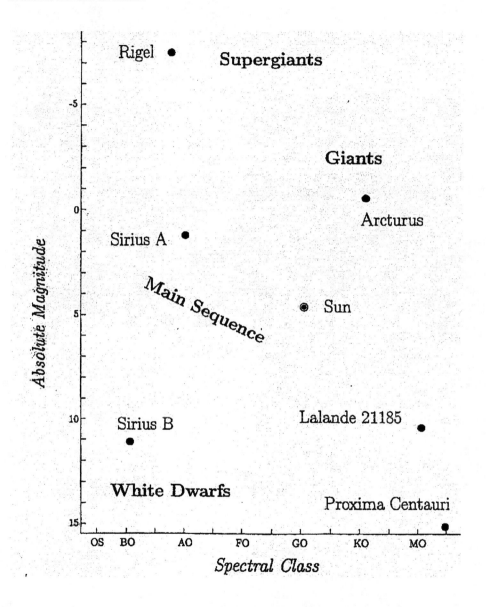

Figure 1.4: The *Hertzprung-Russell* (HR) diagram. The coordinate axes are also proportional to luminosity (visual light with respect to the Sun and surface temperature, instead of the more astronomical terms of absolute magnitude and spectral class respectively.

Regarding our past, it is possible to link brain development and the origin of language with natural selection (cf. Chapter 6). It is remarkable that the growth of the brain of our ancestors took place in the relatively short time of a few million years. Indeed, the australopithecine, who lived 3-4 Myr BP, had a cranial capacity just under 400 cm^3. Today the brain of *H. sapiens sapiens* is more than 300% of that capacity. Furthermore, according to the estimate of the previous paragraph, our brain will still be subject to evolutionary pressures for a very long time. In fact, the normal evolution of the Sun will allow life on Earth for a considerable time: life may persist for over a thousand times as long as the period in which natural selection brought the primitive brain of the earliest hominids to that of man.

Science is at present unable to extrapolate the previous growth of the human brain into the future, but this has not deterred fiction writers from a model for a future speciation event. For instance, the English science-fiction writer Herbert George Wells imagined future speciation events of humanity, which he called the *eloi* and *morlock* species of the genus *Homo* [28].

But, returning to the facts, not fiction, at present, from a combined consideration of cosmic, stellar and biological evolution, it is not clear how contemporary humans will continue to evolve. The role natural selection will play in the future of humanity is no longer clear. As we mentioned above, life on Earth is expected to continue for not more than four billion years. (Once again, we have to emphasize that the Earth will remain only for a finite period of time within the habitable zone of the solar system, since the Sun will continue its own evolution, according to what we know about stellar evolution, as explained above when we discussed the HR diagram.)

Up to the present, however, the role of natural selection has been easier to identify, since for the major part of the existence of our species, technology has not played a significant role in our survival.

Humans, which at one time would not have survived to pass on their genes to progeny, now survive due to an inevitable result of our culture: the development of medicine. It could also be argued that something has changed! Our species is overcrowding planet Earth. The absence of the proliferation of individual species to such a degree was one of the factors that led Charles Darwin to his original insight. The future of humans is still an open problem.

2

From chemical to prebiotic evolution

We have seen in Chapter 1 that the biogenic elements were formed in stellar cores and later were expelled by the host star through stellar explosions [1] and other processes. Subsequently, they combine in the atmospheres of evolved stars to form diatomic and triatomic molecules that are to have transcendental consequences in the subsequent stages of prebiotic and biological evolution. A few examples are: C_2, OH, and H_2O, but even larger molecules have been detected in interstellar clouds.

Organic cosmochemistry.

We now know from observation that, within the range of lower temperatures that occur in circumstellar and interstellar media, some key biogenic molecules are formed, such as hydrogen, ammonia, water, formaldehyde, hydrogen cyanide, cyanacetylene, carbon monoxide and hydrogen sulfide.

The chemistry of interstellar clouds sets the stage for prebiotic evolution. In other words, the nine molecules listed above are sufficient for the synthesis of some of the biomolecules of life, such as the amino acids [1]. It is reasonable to assume that the solar system was formed out of a disk-shaped cloud of gas and dust [2], which we called the solar nebula in Chapter 1. Such disk-shaped clouds have been observed about other young stars.

Most of the original matter of which the solar nebula was made has since been incorporated into its planets and the central star itself, namely, the Sun. Interstellar dust is the product of the condensation of heavy elements, which were themselves produced in stellar interiors, for example, carbon, a major component of interstellar dust, magnesium, silicon and iron.

Later, in interstellar space they combined with elements such as oxygen to form particles measuring typically 0.1 microns. A very significant clue as to the nature of the original solar nebula can still be inferred from the study of comets. In fact, these small bodies formed and remained at the edges of the solar nebula. They represent the primordial conditions of the nebula. We wait anxiously for the Rosetta mission to land on a comet early in the first decade of the present century.

We shall return to this mission again below (cf., also Table 1.1). Yet, as comets pass in the vicinity of the Sun and the Earth, dust particles are released as the comets heat up. Some of these particles may be retrieved in the next few years by the "Stardust" mission (cf., Table 1.1). Such interstellar dust particles (IDP) are a major component of the Galaxy and we find abundant amounts of it in the solar system.

It is not yet possible to distinguish which particles are from comets and which are interstellar dust, hence, the current general interest in the future direct retrieval of cometary matter can be appreciated. Imagine examining one of these primitive particles under a microscope-testing its composition and determining its origin! The average density of particles in space is less than one particle per cubic kilometer. Nevertheless, the total number of particles is large. One can appreciate the ubiquity of these grains by the fact that light is totally absorbed by them before it traverses a distance of 1% of the diameter of the galaxy.

Precursor biomolecules in interplanetary dust

Given the importance of even the simplest molecules that are used by living organisms (biomolecules), it is appropriate to begin with comments on the origin of some of the most important ones. They may have been synthesized in the early Earth, or transported here from space. There are also many indications that the origin of life on Earth may not exclude a strong component of extraterrestrial inventories of the precursor molecules that gave rise to the major biomolecules. About 98% of all matter in the universe is made of hydrogen and helium. In fact, besides hydrogen, the other biogenic elements C, N, O, S and P make up about 1% of cosmic matter. This is a disproportionately large amount if we recall the numerous elements of the periodic table that would be included in the remaining 1% of all cosmic matter.

Some of the more refractory material (i.e., that has gone through chemical reactions at higher temperatures) has remained in the inner solar nebula and has condensed in the form of meteorites, which are called carbonaceous chondrites; these small bodies formed in the early solar system. (Its constituents are silicates, bound water, carbon and organic compounds, including amino acids.) These witnesses to the nature of the solar nebula also contain some of the biogenic elements (cf., Table 2.1).

Some comments on Table 2.1 are needed:

The main lesson to be inferred from the table can be illustrated with the case of oxygen. In spite of being the most abundant element in the Earth's crust (the case of meteorites is similar to that of the Earth), oxygen represents only about 1/1000 of the hydrogen atomic abundance in solar matter.

Cosmochemistry must explain the evolution of the inventories of oxygen and other volatiles on Earth form a solar nebula whose composition is radically different. This is also an open problem in the case of the other terrestrial planets. (Mercury, Venus, the Earth, Mars and, in spite of being the Earth's satellite, the Moon is also included in the group of the terrestrial planets.)

• The chemistry of the solar nebula is the chemistry of the eight elements H, O, C, N, Mg, Si, Fe, and S, and to a lesser extent Al, Ca, Na and P. We leave helium and neon out of this group since they are inert.

TABLE 2.1: Selected abundance of the elements in the solar photosphere and in a carbonaceous (C) chondrite, emphasizing biogenic elements. They are an abundant type of stony meteorite, classified with respect to an analogue-the prototypical Ivuna (I) meteorite. (Chondrites are thus denoted by 2-parameters [3,4].)

Atomic number	Element	Abundance in a CI chondrite $(Si = 10^6\ atoms)$	Abundance in the solar photosphere $(H = 10^{12}\ atoms)$
1	Hydrogen (H)	5.91×10^7	10^{12}
6	Carbon (C)	1.01×10^7	3.98×10^8
7	Nitrogen (N)	3.13×10^6	1×10^8
8	Oxygen (O)	2.38×10^7	10^9
12	Magnesium (Mg)	1.074×10^6	3.80×10^7
14	Silicon (Si)	1×10^6	3.55×10^7
16	Sulfur (S)	5.15×10^5	1.62×10^7
26	Iron (Fe)	9×10^5	3.24×10^7

The abundance of biogenic elements would suggest that the major part of the molecules in the universe would be organic; in fact, out of over a hundred molecules that have been detected, either by microwave or infrared spectroscopy, 75% are organic[5]. Once again, we see an agreement between chemical evolution experiments and observations of the interstellar medium.

Some of the molecular species detected by means of radio astronomy are precisely the same as those shown in the laboratory to be precursor biomolecules. In Table 2.2 we give examples of precursor biomolecules in interplanetary dust particles.

TABLE 2.2: A few precursor biomolecules in interstellar dust particles (adapted from ref. 5).

Molecule	Formula
Hydrogen	H_2
Water	H_2O
Ammonia	NH_3
Carbon monoxide	CO
Formaldehyde	CH_2O
Hydrogen sulfide	H_2S
Hydrogen cyanide	HCN
Cyanacetylene	HC_3N

Cosmic dust and comets: their role in astrobiology

Interstellar dust particles are the predominant form of the condensable elements in the galaxy, which are not in the form of star matter. The evidence supports the view that comets are aggregates of interstellar ice and dust (cf., Fig. 2.1).

In turn, as illustrated in tables 2.3 and 2.4, dust grains are aggregates of inorganic matter (67% of the total mass) and organic matter as well (33% of the total mass). This insight has supported the seminal conjecture that comets are responsible for life on Earth [6]. The grains form in interstellar clouds and in gas outflows from stars. On the other hand, interplanetary dust represents debris recently liberated from comets and asteroids of our own solar system [7]. The results of the chemical analysis is summarized in tables 2.3 and 2.4 below.

Origin of the Solar System: the Rosetta mission

We have seen in this chapter that in our galaxy most of interstellar space is filled with gas in the form of clouds. These formations are evident since their density is higher than the surrounding space. This produces the extinction of light coming to us from distant stars, as well as from distant galaxies. These darker patches of heavens are the birth-place of stars, such as our Sun.

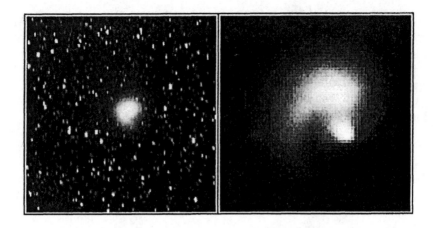

Figure 2.1 The figure shows Comet Hale-Bopp, discovered in 1995 by two independent observers, Alan Hale and Thomas Bopp. The image was taken by the Hubble Space Telescope, and shows material ejected from the rotating comet. The comet was still well outside the orbit of Jupiter (almost 600 million miles from Earth) and due to its size it looked surprisingly bright The closest approach to the Earth was on 22 March 1997 at a distance of 1.3 AU (courtesy of NASA).

TABLE 2.3 : Dust in the Comet Halley: The organic elements (33% of the total mass of the comet).

Elements	Mean chemical composition in mass %
Unsaturated hydrocarbons [8]	16
H, C+ O	5.2
H, C+ N	4.5
H, C+ S	1.8
Water	5.5

The abundance of hydrogen and helium (cf., Table 1.2) is not surprising, since physicists have demonstrated that within the big bang model, elementary subnuclear interactions are capable of generating lighter elements, once the universe is cool enough[9], but it takes the formation of stars, as we have seen, to produce the heavy elements. Before reviewing chemical evolution itself we should first understand current ideas of how the Earth itself originated. The origin of the solar system took place 4.6 Gyr BP by the gravitational collapse of the solar nebula when a certain critical mass was reached.

TABLE 2.4: Dust in the Comet Halley: The inorganic elements (67% of the total mass).

Elements	Mean chemical composition in mass %
Silicates	51.5
FeS (troilite)	6
C (graphite)	3
S (sulfur)	1
Water	5

Thus, the protosun, the protoplanets, the comets, the parent bodies of meteorites, and other planetesimal bodies were formed as the result of this condensation of interstellar matter [10]. The lunar cratering record demonstrates that during its initial stages, the solar system may have been in a chaotic state with frequent collisions of planetesimals amongst themselves as well as with other larger bodies, including the protoplanets. The composition of the solar nebula must have been analogous to that of the interstellar clouds, namely it may have consisted mostly of hydrogen, helium, as well as carbon compounds, dust and ice. This has been confirmed by astronomical studies of Jupiter, Saturn and their satellites.

To get deeper insights into the origin of the solar system we have to wait until 2003for the launching of the above-mentioned Rosetta mission. The name has been taken from the famous "Rosetta Stone", presently at the British Museum in London. This stone has an inscription in three languages, ancient Egyptian hieroglyphs, Demotic Egyptian and Greek. Knowing Greek and having some knowledge of Demotic Egyptian, Champollion was able to use this text to decipher the hieroglyphs. In an analogous manner, the proposed Rosetta mission intends to address inter-related questions: the interstellar medium, cometary material and meteorites. Rosetta will have a lander, for studying the comet nucleus and is expected to rendezvous with comet Wirtanen in 2011 and to land on it in the year 2012.

Origin of the terrestrial planets

From studies of carbonaceous chondrites and the rocky planets, or satellites [11], we know that the terrestrial planets [12] were not formed by a slow process of gradual accumulation of interstellar dust particles that may have fallen in the solar nebula. Thus from the combined information provided by the small objects of the inner solar system and messengers from the outer solar system (comets), we can reconstruct an outline of the main steps in the formation of the terrestrial planets.

We expect high temperatures at the center of the solar nebula and low temperatures at its periphery. This is today an empirical observation for we can observe in our galaxy nebulae where similar processes of star formation are occurring today. As the thin interstellar medium gets concentrated in some regions in space it forms a protonebula, which eventually collapses producing a hot central condensation that becomes the star.

The collapse is coupled with rotation of the gas. A basic law of physics is the conservation of angular momentum. This means that a certain quantity (angular momentum), which depends on mass, distance and rotational velocity is conserved. In the process of collapse the mass remains the same, the distance represented by the dimension of the nebula decreases; so, as a consequence of the conservation law, the angular velocity must increase as the collapse continues. This leads to a proto-Sun with a corresponding rotating disk, which will give rise to the planets and small bodies of the solar system.

Due to the heat, only refractory materials [13], namely materials that vaporize at high temperatures that formed previously in the interstellar space (silicon combining with oxygen into silicate particles is an example [14]) can survive in the inner part of the nebula. This is the first stage of a sequence that will lead to chondrites, planetesimals and eventually to terrestrial planets.

Origin of the jovian planets

By the same physical principle that we referred to in the previous section, namely the conservation of angular momentum, the evolution of the matter distribution of the solar nebula will differ further out form the inner region. Gas is more abundant than IDPs, and yet the temperature will be low enough to allow the existence of water ice

Without going into the details, straightforward physical arguments suggest the reason why the terrestrial planets are more dense than the jovian planets [15]. At a distance of about 5-10 AU the temperature was not low enough to discriminate gases from dust (as it happened in the inner region). Consequently, the composition of Jupiter and Saturn, which were formed at these distances, are expected to reflect the composition of the original solar nebula. These plausibility arguments are subject to predictions that can be confronted with what is known about the chemical composition of the giant planets with that of the Sun. One of the reasons for sending the probe in 1995 into the Jovian atmosphere as soon as the Galileo mission approached the Jupiter system was to put to the test the theoretical foundations of this aspect of planetary science.

Origin of the satellites of the jovian planets

In the complex process of accretion of the jovian planets [16, 17], a further event is relevant for our eventual concern for studying environments where life could arise. When the masses of these planetary cores became large enough through the accretion of solids, they attracted gas from the surrounding solar nebula. It has been estimated that when the core reached about 10 Earth masses or more, the process of adding gas from its surroundings increased rapidly. This led to a situation in which the envelope and planetary core were about equal. Having reached this 'critical' value of the core, further accretion was dominated by the addition of a large amount of external gas.

In our solar system none of the terrestrial planets reached this stage early enough before the gas of the solar nebula was trapped by the Sun or escaped into the outer regions of the nebula or into interstellar space. (Apparently this is not necessarily the case in other solar systems, as we shall see in Chapter 10.)

At the end of accretion, a combination of cooling and the force of gravity led to a process of gradual contraction. During an early part of this phase a circumplanetary disk of gas and dust, in perfect analogy with the parent disk that led to the planets themselves, developed around the jovian planets. The disk originated from the solar nebula as well as from the outer gas envelopes that the planets had already formed.

Satellites formed from these disks, following the same principles that guided the formation of the planets around the Sun itself. For this reason the elements constituting the satellites of the jovian planets would tend to be enriched in those elements that characterized the solar nebula itself; but we have already seen that physical and chemical processes acting in these nebulae can segregate compounds and elements from the solar nebula composition. This remark will be particularly relevant in Chapter 8 when we consider the possibility of life arising in the Jupiter moon Europa.

The observed satellite sizes, distribution and fraction of the mass of the total jovian planet are a clear demonstration that the nebular disk hypothesis works as well for the formation of satellites about planets as it does for the formation of the planets about the Sun.

3

Sources for life's origin: A search for biogenic elements

Until the commitment of the United States by President John F. Kennedy in 1961 to send a man to the Moon by the end of that decade, the only extraterrestrial samples that were available to science were meteorites.

Biogenic elements on the Moon

A principal achievement of the Apollo Program (cf., Table 3.1) was being the first 'sample-return mission', providing Moon rocks that could be tested for the presence of biogenic elements. After more than three decades of that event, the enormous achievement can be seen now in its proper perspective.

The next step in sample return missions will not take place for some years. For various technical reasons the first Mars soil retrieval mission has been postponed till at least the year 2011.

First of all we consider the Moon statistics (cf., Table 3.2). There were no signs of life on our satellite [1], but the question remained whether there had been some degree of chemical evolution over the four billion years since our Moon originated. From 1969 till 1976 samples were returned from nine sites of the Moon surface. (The main biogenic elements that were found are listed in Table 3.3.)

Just a few months before passing away Cyril Ponnamperuma, the director of one of the laboratories chosen for the analysis of Moon samples, spoke during the Third Trieste Conference about his recollections of those very exciting days of the birth of exobiology [2].

The soviet 'Luna Program' was able to retrieve samples from three sites, while the Apollo Program succeeded in six different locations. Many laboratories collaborated in the analysis.

TABLE 3.1: Statistics on the Apollo Program, with emphasis on its aspect as a sample-return mission.

Apollo Mission	Date	Landing site	Weight of the Moon rocks (kgs)
11	20 July, 1969	Mare Tranquilitatis	21.7
12	19 November, 1969	Oceanus Procellarum	34.4
14	31 January, 1971	Fra Mauro	42.9
15	30 July, 1971	Hadley-Apenines	76.8
16	21 April, 1972	Decartes	94.7
17	11 December, 1972	Taurus-Littrow	110.5

Figure 3.1: The Moon as photographed by the crew of Apollo17 in 1972 (courtesy of NASA).

TABLE 3.2: Physical parameters of the Moon.

Parameters	Values
Mass /Earth mass	0.0012
Radius (km)	1738
Eccentricity	0.055
Mean density (g/cm^3)	3.344
Mean Earth density (g/cm^3)	5.520
Period of rotation on its axis, or 'sidereal month' (days)	27.322
Mean distance from the Earth (Earth radii)	60.39
Mean distance from the Earth (km)	384,400

Biogenic elements on asteroids

There are approximately 200 asteroids that have already been identified between the orbits of Mars and Jupiter, where there are many more still to be studied in detail [4]. The largest amongst them is Ceres, some 940 kilometers across (cf., Table 3.4). Perhaps it is necessary to explain the asteroid number shown on Table 3.4. In fact, the first asteroid that was discovered was Ceres, which was discovered 200 years ago, in 1801. It has been labeled as asteroid number 1. At the time of writing a few of them have been imaged by flying spacecraft, for instance, the Galileo mission on its way to Jupiter was able to approach asteroid 243 Ida and its interesting satellite Dactyl (cf., Figure 3.2).

The main interest of asteroids from the point of view of astrobiology is the possible presence of biogenic elements. Collisions amongst asteroids produce smaller pieces that are thrown into orbits intercepted by the Earth. If they survive the transit through our atmosphere these small bodies are called 'meteorites'; they have a chemical composition that can give us a great deal of information on the origin of elements that were to trigger off the process of chemical evolution on Earth and elsewhere (one special body of interest from the point of view of chemical evolution is Titan, cf., Chapter 9). Our knowledge of their composition depends on indirect evidence. Essentially, asteroid composition depends on what we have learnt on meteorites that are available to us (since there is the evidence mentioned above that meteorites do come from the asteroid belt). This information is then combined with results of studying the asteroid properties over the wavelength of reflected electromagnetic radiation, an investigation that is called 'reflectance spectroscopy'. Some relevant meteorites in this context are carbonaceous chondrites (cf., corresponding section below), which have an important biogenic element, namely carbon.

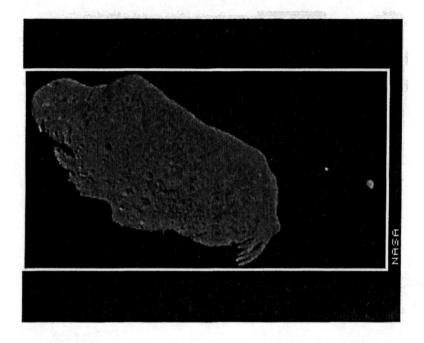

Figure 3.2: The asteroid Ida and its satellite Dactyl (courtesy of NASA).

However, there is another aspect of the chemistry of asteroids that is common to several other small bodies of the solar system, namely satellites, and comets. In fact, since the 1970s various experimental techniques, amongst them infrared observations, have demonstrated that a variety of the small objects contain very dark material which is either black or red in color.

These small bodies are called asteroids of C-type. Normally the term C-type asteroid is reserved for a dark, carbonaceous asteroid in a classification according to the spectra of reflected sunlight. Such asteroids reflect light poorly; we say that their reflectance (or *albedo*) is low. We remind the reader that albedo is defined as the percentage of incoming visible radiation reflected by the surface.

Biogenic elements on comets

Comets are important indicators of the early Solar System due to their orbits as well as to their small size. Evidence has been gathering that some comet nuclei have similar composition to the C-type asteroids (dark carbonaceous and silicate dust). We begin with the small objects that are easily accessible on Earth.

The spectra of several comets are similar (in the light and near ultraviolet range). This suggests that the composition of comet nuclei are similar. The composition contains a great deal of information that has been is brought within reasonable distance form the Earth to allow useful information to be gathered, as it happened towards the end of the 20th century with the Shoemaker-Levy comet collision with Jupiter (cf., Fig. 3.3).

Figure 3.3:   The image shows sites of impact of
fragments of the comet Shoemaker-Levy 9 (courtesy of
NASA/Hubble Space Telescope, 1994).

In fact, the best estimates of their composition based on models and observations
confirm this remark, as illustrated in Table 3.5.

TABLE 3.3: Lunar biogenic element abundance [3] (mean value, parts per million).

Element	Soils	Breccias	Rocks
H	40	68	2.1
C	115	78	1.5
N	82	68	3
S	850	756	1950, basalt. 300, anorthosite.

TABLE 3.4: Examples of some asteroids [4]. (Those with numbers 243, 253 and 951
have been imaged by spacecraft flybys, cf., Fig. 3.2)

Asteroid (name)	Asteroid (number)
Ceres	1
Hygeia	10
Davida	511
Europa	52
Ida	243
Mathilde	253
Gaspra	951

Fred Whipple in 1950 anticipated our current understanding of comet nuclei by modeling. He had already concluded that water ice was the major component. His model is referred to as the "dirty snowball" model.

Biogenic elements on meteorites

We have just seen that between the orbits of Mars and Jupiter there is an important belt of asteroids. In fact, the majority of asteroids in our solar system belong to that belt.

The origin of the asteroids could be found in the nuclei of extinct comets, which themselves may have been formed in the Kuiper belt outside the orbit of Pluto and extending for 500 AU. Others, with longer orbital periods (longer than 200 years, may have been formed in the Oort Cloud which lies beyond the Kuiper belt). Its extension is truly colossal, it goes as far as half-way to the nearest star. An interesting still unanswered question is whether these two largely unknown components of the solar system are present in other solar systems (cf., Chapter 10).

Some of the Apollo family of asteroids may have hit the Earth in the past. Most meteorites are of the stony kind, the so called chondrites, as they contain chondrules (mineral-rich blobs). Amongst the chondrites the most ancient ones, contemporary with the formation of the solar system, are rich in carbon and are, therefore, called carbonaceous chondrites.

There is a very important fraction of all the meteorites that have been collected on the surface of the Earth, whose origin has been estimated to be from the early solar system. They may be referred to as basaltic meteorites. Some important meteorites are tabulated in Table 3.6.

TABLE 3.5: Estimated volatile abundance in comet Hale-Bopp near perihelion [5].

Molecule in comet nuclei	Abundance relative to water
H_2O	100
CO	20
CO_2	6
CH_4 (methane)	0.6
CH_3OH (methyl alcohol)	2
$N H_3$ (ammonia)	0.6
HCN (hydrogen cyanide)	0.2

• THE SNC METEORITES

A large sample of meteorites have been found in Antarctica. From the point of view of astrobiology, the great advantage is that of being far removed from densely populated regions. The low temperatures that are recorded there is a further advantage; this factor helps in the preservation of those valuable extraterrestrial records.

For some time now the composition of the Martian atmosphere has been known, as we demonstrate in Table 3.7.

However, inside some of the meteorites found in Antarctica there is some gas that has been trapped when the minerals were laid down. Such meteorites are called SNC (cf., Table 3.6), an acronym which stands for Shergotty, Nakhla, Chassigny, the name of the places where the meteorites fell.

Studies of these meteorites have shown that the gas that they contained has the same composition as that shown in Table 3.7, hence their origin is assumed to be from Mars. In particular, the Shergotty meteorites are named after the city in India where the prototypical meteorite fell in 1865.

Further support for the Martian origin of the SNC meteorites comes from a comparison of the mineral content of a Shergotty meteorite and that of Martian soil, as we show in Table 3.8. The SNC meteorites were driven off the surface of Mars by large impacts produced by large objects. In time a few of these have been retrieved by geophysicists working in Antarctica.

TABLE 3.6: Some important meteorites. The SNC meteorite will be considered in detail in the next section. The initials S, N and C stand for the locations where these meteorites were first retrieved (shergottites, nackhlites and chassignites, respectively). The Murchison meteorite will also be discussed below in a separate section [6].

Place where the meteorite fell	Type	Weight
Grootfontein, Namibia	Iron (90% Fe, 10% Ni)	60 tons
Jilin, China	Stony	1.7 tons
Murchison, Australia	Carbonaceous chondrite	Many fragments
Allan Hills, Antarctica	SNC meteorite (Martian)	2 kgs

Another important aspect of these meteorites is that their mineral composition implies that they were at one time in water. This is one of the indicators that Mars may have had in its past a 'clement period', with conditions appropriate for the origin of life.

• THE MURCHISON METEORITE

This meteorite has proved to be useful in fundamental problems which concern us in this book. We begin by discussing some of these questions and, later, we shall return to the significant discoveries that have been derived from the Murchison meteorite.

Eukaryogenesis occupies a central position in a wide range of problems in astrobiology (cf., Chapters 6, 8 and 12), ranging from chemical evolution to bioastronomy. The identification of eukaryogenesis is, in turn, relevant to another major problem, SETI (cf., Chapter 11), since intelligent behavior can occur in multicellular organisms of eukaryotic cells. In Chapter 12 we argue that a cellular plan different from the eukaryotic one is unlikely to lead to multicellular organisms with intelligent behavior.

TABLE 3.7: Composition of the Martian atmosphere.

Chemical component	Fraction by weight (%)
Carbon dioxide, CO_2	95
Molecular nitrogen, N_2	1.5
Water	0.03

TABLE 3.8: Minerals in shergottites and Martian soil[6].

Mineral	Formula
Silicon dioxide	SiO_2
Ferrous oxide	FeO
Calcium oxide	CaO
Magnesium oxide	MgO
Aluminum oxide	Al_2O_3
Titanium dioxide	TiO_2
Potassium oxide	K_2O

In addition to the empirical SETI approach, exploring the consequences of the laws of physics and chemistry may give us some insights on the question whether or not we are alone in the cosmos. The concept of the constraint on chance (cf., Chapter 5), militates against the older criterion of Monod who, during his lifetime, had less information than we have accumulated now on the mechanisms that are behind the origin and evolution of life on Earth.

Indeed, chance (rate of mutations) and necessity (natural selection) do not imply that life elsewhere in the cosmos is unlikely [7] Thus, the prospect for detecting life in the newly discovered planets (cf., Table 10.1) remains a possibility, either by detecting life-supporting volatile elements, or by directly detecting their radio messages.

It has been clear for some time now that the extraterrestrial option should not be ruled out. Firstly, a significant illustration of the plausibility of an extraterrestrial origin of the precursors of the biomolecules is based on the meteorite which fell in the town of Murchison in south-eastern Australia on September 28, 1969. Since the meteorite exploded in mid-air, many fragments were retrieved near the town. This meteorite is classified as a CM2 meteorite, as carbonaceous chondrites are classified into 9 classes according to a two-parameter system as follows:

• Ivuna,

• Mighei,

• Vigarano,

• Ornans.

(There are five more classes of chondrites which are not referred to in the present work; the reader should notice that the abbreviations CI, CM, CV, CO refer to the analogues of the prototypical meteorites listed above.)

The laboratory of Cyril Ponnamperuma was able to obtain a piece of the Murchison meteorite. At that time they were preparing for the first analysis of the lunar rocks. They were able to get the first conclusive evidence of extraterrestrial amino acids. The results obtained by Ponnamperuma's group and others have demonstrated the universality of the formation of some organic compounds which are essential for life today [2]. The result of the analysis is shown in Table 3.9.

However, the above chemical evolution sketch of the origin of life is incomplete, a fact that is illustrated, for instance, with the question of the gases that were present in the early Earth atmosphere, a topic which is not settled [11]. Perhaps one of the most important aspects of the analysis of the Murchison meteorite is the fact that in that meteorite a detailed analysis of the relative abundance of the precursors of the biomolecules is known. This is illustrated in Table 3.10.

On the other hand, carbon dioxide must have been sufficiently abundant to have prevented the Archean Earth from freezing under the Sun at a lower level in the main sequence of the Hertzsprung-Russell (HR) diagram. Without reference to the HR diagram, in simpler language we may say that at the time that life first appeared on Earth, the Sun was some 30% less luminous than it is today. This is referred to as the *'faint young Sun paradox'*.

Through a greenhouse effect there would have been the appropriate temperatures for producing the bacterial associations that we know must have been in existence some 3.5 Gyr BP (cf., Chapter 2). Besides, it also remains to be clarified what was the relative importance of the precursors of life that were brought to Earth by comets, meteorites, and micrometeorites compared with the inventories that were part of the Earth as it formed out of our own solar nebula.

• THE ALLAN HILLS METEORITE

Another meteorite, which is relevant for exobiology has been retrieved in 1984 by a National Science Foundation mission from the wastelands of Antarctica in a field of ice called the Allan Hills (cf., Fig. 3.4).

The meteorite weighs about two kilograms. The presence of an important biomarker has been pointed out by David McKay and co-workers at NASA [12], namely, the polycyclic aromatic hydrocarbons (PAHs). On Earth PAHs are abundant as fossil molecules in ancient sedimentary rocks, and as components of petroleum.

The presence of PAHs in the ALH84001 meteorite is compatible with the existence of past life on Mars. This result, which requires confirmation, underlines the importance in the future of formulating comprehensive questions (as suggested in this section), regarding life on Mars, to include *the degree of biological evolution* covering the complete range known to us, from simple bacteria to the eukaryotes.

TABLE 3.9: Amino acids from an extraterrestrial source
which have also been obtained in experiments [8,9].

Amino acids in the Murchison meteorite, which occur in similar relative abundance as in experiments	*Amino acids which occur in more relative abundance in the Murchison meteorite than in experiments*
Glycine	Valine
Alanine	Proline
Aspartic acid	Glutamic acid

In this context, we should underline that from the experience that we have gathered on Earth [13] we know of at least two important groups of abundantly distributed biomarkers which have been characterized from oils and sediments (the fossil triterpanes and steranes), whose parental materials are found *almost exclusively in eukaryotes.*

Therefore, if the degree of biological evolution in Mars has gone as far as eukaryogenesis (or taken the first steps towards multicellularity) then, there are means available for detecting this important aspect of exobiology in future space missions (i.e., the search for the parental materials for steranes: tetracyclic steroids).

Biogenic elements in extreme terrestrial environments

In most habitats the available energy is due to the Sun; in other words, the food chain begins when plants trap the energy they need by means of the process of photosynthesis.

Both plants and some bacteria convert energy from the Sun into chemical energy, which is used to produce carbohydrates, a class of organic compounds of great biological importance both structurally and as energy stores.

However, there are alternative energy sources to the Sun at the bottom of the oceans and other water environments, including the Gulf of Mexico which harbors a rich fauna supported by underlying fields of oil and gas. The vent environments are fuelled mainly by geologic processes including volcanic activity. We illustrate this point with hot springs.

However, there are alternative energy sources to the Sun at the bottom of the oceans and other water environments, including the Gulf of Mexico which harbors a rich fauna supported by underlying fields of oil and gas. The vent environments are fuelled mainly by geologic processes including volcanic activity. We illustrate this point with hot springs.

Figure 3.4: The Allan Hills meteorite ALH84001,0 The small black box on the left represents a scale of 1 cm (courtesy of NASA).

Indeed, ever since deep-sea hydrothermal vents were discovered in 1977, they have been found subsequently in several tectonically active areas in the ocean floor. John Corliss and others have been defending the thesis that life may have originated at hydrothermal vents. This conjecture raises an interesting point regarding our understanding of the possible distribution of life in the solar system. In fact, with our present incomplete understanding of life's origins, it is not at all evident that living organisms will evolve exclusively within, or close to what has traditionally known as a "habitable zone", namely the region that is neither too near (and hence too hot, i.e., Mercury) nor too far from the Sun (and hence too cool, i.e., Jupiter), in order to allow life to emerge. Effectively the complete extension of the Solar System is a possible site for the evolution of life, if the hypothesis that life could emerge at hydrothermal vents is correct. This hypothesis, however, must await complete experimental confirmation, a point to which we shall return in Chapter 8, when we discuss the possibility that microscopic life could exist in the Jovian satellite Europa. Mars, the most promising

We shall return to this question later on.

TABLE 3.10: Selected precursors of the biomolecules in the Murchison meteorite. (The total carbon content is 2.0-2.5%.). The abbreviation PPM denotes parts per million [8].

Precursors of the key biomolecules	*Abundance*
Carbonate and CO_2	0.1-0.5 %
Hydrocarbons	
aliphatic	12-35 PPM
aromatic	15-28 PPM
Amino acids	10-20 PPM
N-heterocycles	
purines	~ 1 PPM
pyrimidines	~ 0.05 PPM

On the other hand, current parallel evolution based on a totally primary production based on chemosynthetic bacteria cannot be excluded from the outer solar system satellites, where we know that there exists frozen water, or water in hydrated silicates. As we mentioned above, the most prominent candidate is the Jovian satellite Europa.

• HOT SPRINGS AND THE SEA FLOOR

Close to the warm vents there exist dense invertebrate communities, in which the chemosynthetic bacteria are in symbiosis with organisms in the ecosystem. For instance, bivalves and gastropods live in symbiosis with their bacteria in their gill tissues. Vestimentiferans, polychaetous annelids have been observed with a very broad distribution throughout the ocean floor.

In the bottom of the Gulf of Mexico, over 500 meters from the surface there is an underwater lake of brine so dense that it remains in a depression of the sea floor. A vast salt deposit produces the gradual seeping through of gases that autotrophic bacteria can turn into foodstuff, and these bacteria become the first stage in the food chain of clams, mussels and tube worms. Because they are so foreign to our everyday experience and also far removed by catastrophes that may exterminate other ecosystems, the hot-spring environments have been assumed to be a sort of refuge against evolution, as it is the background of the elimination of species that new species evolve. It has also been observed by *in situ* submarine investigation that hot-spring communities of animals are remarkably similar throughout the world.

This 'refuge' concept has prevailed even in astrobiology: There has been some speculation on the possible origin of life at the bottom of the Europan putative ocean, who have only considered that archaebacteria may have also evolved in the benthic regions (deep sea), which were heavily dependent on bacterial chemosynthesis.

However, scientists at the Urals branch of the Russian Academy have identified fossils from the earliest hydrothermal-vent community dating from the late Silurian Period over 400 Myr BP. This particular community has its own case of species extinction, as the fossils have been identified and correspond to lamp shells (inarticulate brachiopods) and snail-like organisms (monoplacophorans).

This discovery has the profound implications when specific biological experiments shall be designed for testing for the presence of life in Europa, as we cannot maintain today that the analogous conditions that may exist there will induce the appearance of archaebacteria-like organisms.

Given the common origin of all the Solar System, the antiquity of the favorable conditions in Europa may have been conducive not only to the first steps in evolution. On the other hand, the source itself of life (hydrothermal vents at the bottom of the Europan ocean) cannot be considered a permanent refuge for only archaebacteria, but the most transcendental transition into a complex cell, possibly a eukaryote, has to first be ruled out by experimental tests. The effort is worth taking, as it was the most momentous step in the evolution towards intelligent behavior on Earth. The theoretical basis arguing in favor of common basic cellular plans for life in the universe have been argued in the past. The experimental proposal for a space mission has been proposed.

Finally, we shall argue that the experimental tests are not beyond the present capacity of miniaturization of the traditional microscopic fluorescence that may distinguish between prokaryotes and eukaryotes. Simpler tests are possible, for instance, search for a mitotic division or even meiosis; but these simpler tests are not satisfactory, as we must aim at answering first the most basic differences at the simplest stage of eukaryotic evolution. The case of the primitive eukaryote *Cyanidium caldarium*, suggests that distinguishing the degree of evolution of a given microorganism, is not straightforward in the context of experiments that should be performed *in situ* in either a planet such as Mars, or even a satellite such as Europa.

Perennially ice-covered lakes in Antarctica

A terrestrial analogue for frozen satellites, in which there may be volcanic activity that contributes to melting the ice under a surface that remains permanently frozen is found in a series of dry valleys discovered by the British explorer Sir Robert Scott in 1905. They are near the American Base at McMurdo in Antarctica. Some of the most interesting lakes in that region are permanently covered by ice.

From the point of view of geology and biology some of the best studied frozen lakes are in the Taylor Valley (bounded by the Ferrar Glacier and the Asgard Range), namely Lake Fryxell and Lake Hoare; further north, in the Wright Valley Lake Vanda is also remarkable. In Table 3.11 we give some details of a few of the lakes of the Dry Valleys. Amongst the microorganisms that are permanently living in the frozen lakes there are examples of both prokaryotes as well as eukaryotes Besides some of the most interesting geologic paleoindicators for reconstructing the history of these lakes are organosedimentary structures known as stromatolites formed by various species of cyanobacteria, such as *Phormidium frigidum*.

In this sense the existence of these permanently frozen lakes add an extra bonus to our working model of the Europan Ocean. Modern organisms analogous to the early biota of the Earth are to be found in the Dry Valley lakes. What is most important, single celled eukaryotes are amply represented in this biota. Amongst the paleoindicators that have been found are diatom frustules, cyst-like structures, most likely of crysophycean origin have also been identified. In Tables 3.12 and 3.13 we summarize some of the organisms known to inhabit in these lakes.

Lake Vostok

From the point of view of the possibility of the existence of life on Europa (cf., Chapter 8), we should consider a lake called Vostok, which is the largest of about 80 subglacial lakes in Antarctica. Its surface is of approximately 14,000 km^2 and its volume is 1,800 km^3. Indeed this Ontario-sized lake in Eastern Antarctica is also deep, with a maximum depth of 670 m. On the other hand, from the point of view of microbiology, the habitat-analogue provided by Lake Vostok for the Europa environment seems appropriate. At the time of writing the ice above the lake has been cored to a depth of over 3,600 m, stopping just over 100 m over the surface of the lake itself. This work has revealed great diversity of single-celled organisms: yeast, actinomycetes, mycelian fungi (which remain viable for almost 40,000 years), the alga *Crucigenia tetrapodia,* diatoms, and most interestingly, 200,000 year old bacteria. Besides it appears that water temperatures do not drop too far below zero centigrade, with the possibility of geothermal heating raising the temperatures above this level.

Extrapolation of data retrieved from work deep in the ice core to the lake itself, implies that Lake Vostok may support a microbial population, in spite of the fact that that large volume of water has been isolated form the atmosphere for over one million years [19]. The reader will appreciate the relevance of this work after reading Chapter 8, which is devoted to the jovian satellite Europa. Indeed, we shall discuss the possibility of the existence of Europan microbes within a liquid water environment sealed by a frozen surface, which is by now known in great detail with the help of the Galileo Mission (cf., Fig. 8.1).

TABLE 3.11: Statistics of the Dry Valleys lakes in Antarctica (After refs. [14-16]).

Lake or pond	Maximum depth (meters)	Elevation (meters above sea level)	Lake type
Lake Fryxell	18	17	Perennial ice cover; liquid water
Lake Hoare	34	73	Perennial ice cover; liquid water
Lake Vanda	69	123	Perennial ice cover; liquid water
Don Juan Pond	0.1 - 2	116	Ice-free

TABLE 3.12: Microorganisms living in the Dry Valleys lakes, Antarctica [14-18].

Organism	Domain	Habitat
Cyanobacteria	Bacteria	Lakes Chad, Fryxell and Vanda
Leptothrix	Bacteria	Lakes Fryxell and Hoare
Achronema	Bacteria	Lakes Fryxell and Hoare
Clostridium	Bacteria	Lakes Fryxell and Hoare
Chlamydomonas subcaudata (Phylum Chlorophyta)	Eucarya	Lakes Bonney (east lobe) and Hoare
Diatoms (Phylum Bacillariophyta)	Eucarya	Lakes Bonney, Chad, Fryxell, Hoare and Vanda
Bryum cf. algens (a moss)	Eucarya	Lake Vanda

TABLE 3.13: A few examples of eukaryotes present in Antarctica [14-18].

Organism	Domain	Habitat
Diatom shells	Eucarya (Bacillariophyta)	Lake Vostok (ice core, at depth of 2375m)
Caloneis ventricosa	Eucarya (Bacillariophyta)	Lakes Chad, Fryxell, Hoare and Vanda
Hantzschia amphioxys	Eucarya (Bacillariophyta)	Lakes Fryxell, Hoare and Vanda
Navicula cryptocephala	Eucarya (Bacillariophyta)	Lakes Bonney, Fryxell, Hoare and Vanda
Chlamydomonas subcaudata	Eucarya (Chlorophyta)	Lakes Bonney and Hoare
Tetracystis sp.	Eucarya (Chlorophyta)	Lakes Fryxell, Hoare and Vanda
Yeast	Eucarya (Ascomycota)	Lake Vostok (ice core)

Part II

PREBIOTIC EVOLUTION:
THE BIRTH OF BIOMOLECULES

4

From prebiotic evolution to single cells

In our studies of the origin of life we will encounter the major macromolecules of life: proteins, nucleic acids, polysaccharides and lipids. We shall learn that there is great unity in all of biochemistry. This important remark will be illustrated with two stunning examples: the analogous asymmetry of all the main molecules of life and secondly, the universality of the genetic code. [More often than elsewhere in the book the reader may find it necessary to refer to the notes and the glossary for a deeper appreciation of the concepts discussed in this chapter.]

Which are the macromolecules of life?

To understand the next great transition in the ascent of life from organic chemistry to the living cell, we must first comment on certain molecules that have played a key role in that ascent.

In some cases we will be considering large macromolecules that take part in the all-important process of producing multiples copies of themselves, in order to drive the living process. They are, on the one hand, amino acids and the polymers they form, namely the proteins. On the other hand, we have the bases and the polymers they form, the so called nucleic acids.

Later on, when we consider our earliest common ancestor, we shall have to refer to other molecules of life that serve to cover up into a sac-like structure the proteins and nucleic acids already mentioned that drive the basic process of life, that of information storage (in the nucleic acids) and the translation of such information into useful proteins (the combined action of nucleic acids and proteins). In other words, a third major group of macromolecules that are relevant to life consists of those molecules, the lipids, which due to their response when they are in contact with water have become essential in the formation of membranes around the cells that contain proteins and nucleic acids.

Let us consider the main molecules that we have already referred to above:

• Amino acids are any of twenty organic compounds that are the building blocks of proteins, which are synthesized at one of many small cellular bodies called ribosomes. In meteorites over 70 amino acids have been detected, but the proteins of living organisms only make use of 20 amino acids. Their three-letter abbreviations are given in Table 4.1. (The precise chemical formula of the amino acids is not essential for following the arguments in this book.)

TABLE 4.1: The 20 amino acids and their three-letter abbreviations.

Ala: Alanine	Gly: Glycine	Pro: Proline
Arg: Arginine	His: Histidine	Ser: Serine
Asn: Asparagine	Ile: Isoleucine	Thr: Threonine
Asp: Aspartic acid	Leu: Leucine	Try: Tryptophan
Cys: Cysteine	Lys: Lysine	Tyr: Tyrosine
Gln: Glutamine	Met: Methionine	Val: Valine
Glu: Glutamic acid	Phe: Phenylalanine	

• Proteins are organic compounds, which are essential biomolecules of all living organisms. Their elements are: hydrogen, carbon, oxygen, nitrogen and sulfur. They are made up of a series of amino acids. (A medium-sized protein may contain 600 amino acids.)
    • The nitrogenous base compounds U, C, G, A and T are ring compounds which are constituents of nucleic acids. These letters represent the initial of the corresponding base; for instance, U denotes uracil. The full name for the other bases are: cytosine, guanine, adenine and thymine. To follow the arguments presented in this book the detailed chemical formulae are unnecessary.
    • Nucleic acids are organic molecular structures consisting of five-carbon sugars, a phosphate and mainly, although not exclusively, one of the five bases already mentioned above.
    • Lipids are a wide group of organic compounds having in common their solubility in organic solvents, such as alcohol. They are important in biology, as they are constituents of the cell membrane and have a multitude of other important roles.

The primitive Earth

There are many reasons why we cannot be certain of the exact conditions in which life evolved on Earth. One important factor is what has been called the heavy bombardment period in the early solar system. Wherever conditions may have prevailed in the atmosphere, oceans and lithosphere, they must have been altered significantly when a large number of large bodies collided with the early Earth. So much so that the Moon itself is believed to have been the product of a massive collision of the Earth and a Mars-like object. Entire oceans may have boiled off down to a depth of several kilometers.

Nevertheless, there is another body in our Solar System which is within reach of experimental probing. Indeed, there is already a mission on its way to Titan, the satellite of Saturn. We shall discuss this in detail in Chapter 9; suffice it to say at the moment that Titan's atmosphere is dense and lacking in oxygen. This satellite provides us with a scenario analogous to the early Earth, giving new meaning to the earliest experiments searching for the origin of the biomolecules of life, which have already been enumerated in the previous section: the amino acids, nucleic acids and lipids.

Stanley Miller, after some significant previous experiments by Melvin Calvin, attempted to create a model for the early Earth, in which it was postulated that the main components of the atmosphere were methane (as in the case of Titan), ammonia, hydrogen and water (cf., also Chapter 15). His experimental set up included a flask of water which was boiled to induce circulation of the gases, which at the same time was capable of trapping any volatile water-soluble products which were formed during the experiment. An electric spark acted on the gases for a period of time. After suspending the electric discharge the water was found to contain several small organic compounds, two of which are found in all proteins: the amino acids glycine and alanine.

Since the Miller experiment was concluded half way through last century, the debate has continued regarding the nature of the primitive atmosphere. Today, some arguments lead us to think that the original gases may not have coincided with those of the Miller experiment. However, that is not the main point of the revolutionary trend that Miller started. What is more significant is that oxygen was missing. If Miller had been advised by his tutor, the Chemistry Nobel Laureate Harold Urey, to add oxygen to his gas mixture no amino acids would have formed. This will be rationalized later on in this book, when we shall mention that the generation of oxygen was mainly due to life itself.

The origin of the first cell

We are about to enter the most exciting phase of our ascent form the cosmic creation of the basic molecules of life to the stage of encapsulating them into a living cell with a membrane of lipids. In the interior of the cell the chemistry of life will take place.

We still do not have the final answer about the origin of the first cell, although much research has gone in this direction. Two proposals have been associated with the earliest work in the area. The Russian chemist Alexander Oparin suggested that a mixture of organic polymers when heated can assemble themselves into a membrane, which in principle could contain the rest of the constituents of the living cell; since this early twentieth century contribution was published, the Oparin structures - the coacervate droplets - can properly be said that they could contain the genetic material, the biopolymers DNA and RNA made up out of the four bases that were mentioned above. (The T base, which is a constituent of DNA, is systematically replaced in RNA with a U base.)

Much impressed by this work, an American scientist, Sidney Fox suggested that small, spherical membranes can assemble spontaneously by heating amino acids in water. Fox called these structures 'proteinoid microspheres'. He persisted with his work for half a century with numerous colleagues. We have a wonderful memory of Sidney

Fox's lecture in Trieste where he delivered a lecture co-authored by twenty colleagues who collaborated with him over a good fraction of his career.

The essential contribution of Oparin, Fox and others was to demonstrate that the entrapment of biomonomers and biopolymers within a membrane was possible in conditions that may have resembled those of the early Earth. However, it is also true that it is not evident how these proto-membranes, the forerunners of the modern cell, may have given rise to the membranes that we see today.

Independent experiments by a group of younger scientists that began their work after Fox, have demonstrated that under primitive Earth-like conditions the real components of the modern cell, the lipids, may indeed be synthesized. Joan Oro, Cyril Ponnamperuma and many other organic chemists have led the way to a more comprehensive picture.

A biochemical relic of the earliest stages of life

Independent of the exact details of the pathway that may have led to life on Earth, today all living organisms have extraordinary relics of the early stages of life. We shall complete this necessarily sketchy chapter by focusing attention on two of them. The simplest is a universal handedness that is observed from bacteria to humans, a certain asymmetry of all the biomolecules, although for simplicity we shall refer to amino acids. The second will be an almost universal code that translates sequences of genes into proteins.

In order to do so we must first cover a certain number of physical and chemical properties of matter in general. Indeed, many chemical substances, when they are extracted form a liquid environment, in which they are dissolved, assume a definite crystalline form.

Generally, crystals are regular in form. For instance, they may be divided by a plane into two symmetrical halves. But this is not always the case. Sometimes there is an asymmetry which has been understood since the time of the great French scientist Louis Pasteur.

A most remarkable aspect of the origin of life on Earth is the unity of biochemistry. At the lowest level an asymmetry makes that unity evident. The key biomolecules have the same 'handedness'. This phenomenon occurs when molecules are asymmetric, in such a manner that they are able to exist in two configurations, which like a pair of gloves, mirror each other's shape; this is referred to by saying that both partners (or 'stereo-isomers') are mirror images of each other, or that such stereo-isomers have the same handedness. For example, amino acids all come in two versions, which are technically referred to as 'two stereo-isomers'.

These molecules are optically active, just like many three-dimensional structures. As we said above, we owe to Pasteur our understanding of the relationship between molecular asymmetry and response to the light going through a liquid in which they are dissolved. Pasteur in 1848 devoted his studies to the effect on light of mixtures of certain crystals.

Thus, the key macromolecules of life have the same handedness, or, more often these molecules are said to have the same 'chirality' or, preferably that they are 'homochiral'. (These words are taken from the Greek language, as *cheir* means hand.) Finally, a mixture of equal quantities of the left- and right-handed forms of an optically active compound is called a 'racemic mixture'. In simpler terms, when amino acids are created in the laboratory by means of a chiral device, they are said to be racemic (they contain equal numbers of left- and right-handed molecules).

   Chiral molecules have non-superposable three-dimensional mirror image structures or 'enantiomers' (once again, these are words derived from the Greek *enantios morphe*, whose meaning is 'opposite shape'). Molecules that respond to beams of light in the above-mentioned manner are said to be optically active. An example of single-handed molecules that will concern us in this review are the monomers of proteins, which are surprisingly and exclusively left-handed amino acids.

The origin of the macromolecules of life

In Table 4.2 we indicate some of the chemical reactions that lead from the precursors to the macromolecules of life themselves. Altogether, about 12 molecular species that occur spontaneously in interstellar clouds have been shown to be precursors of the main biomolecules. Table 4.2 is a summary of research that began in the 1950s and extends right to the present, but will undoubtedly continue its robust progress in the future. Comprehensive reviews are given in the Trieste Series on Chemical Evolution (cf., Introduction, refs. 15-20).

Extraordinary symmetry of cell membrane molecules

Phospholipids are biomolecules of the cell membrane having a stable ion formed from phosphoric acid H_3PO_4 (a phosphate group) and one or more molecules with two distinct regions reacting differently with water, one highly soluble, called hydrophilic, the other being water insoluble fatty acids. For this reason such molecules readily form lipid bilayers that are the basis of cell membranes.

TABLE 4.2 : Reactions relevant to chemical evolution [2].

Precursor molecule	*Macromolecule of life*
Formaldehyde CH_2O	Ribose, glycerol
Carbon monoxide + hydrogen $CO + H_2$	Fatty acids
Hydrogen cyanide HCN	Purines (adenine, guanine)
Cyanamide H_2NCN	Peptides, and phospholipids

These are also 'chiral' molecules in the same sense as amino acids, as we explained above. We may say that they have biomarkers related with their handedness, as we shall proceed to explain. In phospholipids two of the -OH groups of a constituent molecule (glycerol) are linked to fatty acids. In fact, glycerol is a colorless, sweet-tasting viscous liquid, widely distributed in all living organisms. (At the molecular level its atomic formula is that of an alcohol.) Indeed, glycerol illustrates a very important property of the cell membrane of many microbes. We will see repeatedly that at the cellular level we may group microorganisms into three large groups or taxons called domains; in fact, the first two taxons contain exclusively right-handed-glycerol [3]:

Firstly, Bacteria is the group or 'domain' of all true bacteria, the so-called 'eubacteria'. Secondly, Eucarya is the domain which includes animals, ciliates, green plants, fungi, flagellates and microsporidia. Finally, archaeans are exceptional microorganisms which have left-handed glycerol in the phospholipids of their cell membranes. They were previously called archaebacteria and form the third group of all life on Earth (Archaea). These microorganisms live in extreme conditions with respect to temperature, pressure, or salinity. From the point of view of molecular biology we shall return to this aspect of modern taxonomy towards the end of this chapter.

Are there extraterrestrial sources of molecular asymmetry?

Molecular asymmetry ('homochirality') may be considered as a signature for life anywhere in the universe. Due to the progress in research and the availability of a whole series of forthcoming space missions currently being planned beyond the first decade of this century, the problem of chemical evolution has been virtually transferred from the laboratories of organic chemistry to the space sciences.

In this context we may recall intriguing ideas which suggest, independently, that circularly polarized light from neutron stars on passing clouds of interstellar dust may selectively eliminate one or the other enantiomer in the dust mantle [4,] if mirror image molecules are originally present [5,6]. (Circular polarization of a light wave refers to the orientation of its oscillating electric field, which rotates 360 degrees clockwise or anticlockwise during each cycle.). According to some recent work another possible astrophysical source of homochirality may be supernovae [7.]

The discovery in 1969 of the Murchison meteorite (cf., Chapter 3) was a landmark in our understanding of terrestrial macromolecules. For Murchison contained non-biological amino acids, indeed not present in the proteins of terrestrial organisms. In fact, of the amino acids that have been identified in the Murchison meteorite, about 50 of them are non-protein amino acids[8.] These numbers give us a considerable insight into the origin of life, when we recall that the protein constituents are chosen from a very restricted set of twenty amino acids.

Was there an asymmetric influence in chemical evolution?

Analysis of the Murchison meteorite has been notoriously difficult ever since it fell in Australia, while the main chemical evolution laboratories around the world were preparing themselves for the analysis of the Moon samples returned by the Apollo 11 mission [9]. All the preliminary experiments searching for a chiral bias reported negative results.

However, it has been observed that the endogenous amino acids contained in the meteorite coincide in part with those that constitute proteins. There is a direct approach to detect deviations of racemisation (i.e., deviations from equal numbers of left-handed and right-handed amino acids). With this objective only those amino acids that are not biogenic have been studied, including those that are rare on Earth.

In this manner, the difficulty is removed of detecting a chiral bias, as well as the uncertainty of whether the positive result is due to biogenic contamination. Following this strategy, some rare amino acids in the Murchison meteorite have been shown to have a small excess of left-handed versions of four amino acids ranging up to 10% [10].

The handedness of biomolecules are useful biomarkers

Homochirality of the macromolecules of life is valid for all organisms, but some care most be taken with the concept of homochirality as we consider the highest taxa:

Exceptionally, in the domain Bacteria (encompassing the flavobacteria and relatives, the cyanobacteria, the purple bacteria, the Gram-positive bacteria, and the green non sulfur bacteria), cell walls may contain right-handed amino acids, as in the case mentioned in the Introduction of *Lactobacillus arabinosus*.

A second exception should be stressed. It concerns the domain Archaea, which, for instance, includes the genus *Pyrodictium* and the genus *Thermoproteus;* such archaebacteria are capable of producing methane as a by-product of the reduction of carbon dioxide. These microorganisms are known to contain left-handed-glycerol in their membrane phospholipids, instead of the standard right-handed-glycerol as we mentioned above, which is characteristic of the phospholipids of the cellular membranes of the other two domains, Eucarya and Bacteria.

The differences that we have pointed out in this section are relevant, since it is important at all stages in the study of the origin and evolution of life on Earth to be aware that we can recognize the degree of evolution of microorganisms by noticing wherever possible the 'biomarkers' that are characteristic of each of the highest classification groups (or taxons).

For example, evolution from the simple archaebacteria with characteristic left-handed glycerol in their cell membranes differ from the more evolved nucleated cells of higher organisms that have right-handed glycerols in their cell-membrane phospholipids.

In the future campaign of exploration of the solar system, it is necessary to distinguish the degree of evolution of putative living microorganisms in environments that may be favorable to life.

An analogy with languages: the genetic code

The analogy with languages will allow us to make a comfortable summary of the major facts that have been learnt regarding another major insight in our understanding of the origin of life.

We are referring to the language of proteins and nucleic acids. It concerns the establishment of a 'genetic code'. This may be considered as a dictionary, which translates from the language of nucleic acids (i.e., DNA and RNA) to the language of proteins. Both of these classes of biomolecules are basic for all the living cells on Earth.

It is the bases and amino acids that constitute the genetic alphabet. The language of the DNA consists of words of four letters corresponding to the four bases of the nucleic acids that we have mentioned above (T, C, A, G). The language of the proteins consists of words of 20 letters corresponding to the 20 amino acids (cf., Table 4.1) that take part in the structure of proteins.

Nucleic acids, in particular RNA molecules, are capable of serving as a first step in the implementation of the transfer of the information. This transfer takes place from triplets of bases that code for a given amino acid (for this reason they are called 'codons'), according to the standard genetic code (cf., Table 4.3).

In other words, since U is a constituent base of RNA and T replaces U as a constituent base of DNA, only four letters of the (RNA) nucleic acid dictionary (U, C, A and G) are used in the genetic code.

On the other hand the codon denoted as "stop" needs some explanation: It is sometimes called a *termination codon* and serves, within the genetic machinery, to signal the cutting of the nascent protein that is in the process of synthesis. In other words, the stop codon indicates that the corresponding triplet terminates the polypeptide chain that is being synthesized on any of the ribosomes [11].

Two examples should suffice to grasp how the code is read and interpreted: To begin with, consider the bottom right-hand corner of the code shown in Table 4.3:

First of all, we learn that both 'codons' GGU and GGC code for the amino acid glycine. (The reader should refer to Table 4.1 for the amino acid abbreviations.) Glycine can also be coded by another two codons.

Secondly, we can think of phenylalanine as being coded by the presence of the bases UUU and UUC on the RNA molecule that codifies the genetic information.

The carrier of the genetic information is a transcript of the original DNA, which is an RNA molecule. This molecule has the capacity of travelling from the nucleus to the site where the synthesis will be carried out, namely, the ribosome. For this reason it is given the reasonable name of "messenger-RNA", and is abbreviated as mRNA. A set of enzymes is responsible on the ribosome for the joining of the amino acids into the full protein. The reader will appreciate at this stage the enormous step that the discovery of the genetic code meant for the science of molecular biology. For the first time one was able to comprehend how the sequence of bases on the DNA that carries the genetic information is related with the sequence of amino acids in protein synthesis. Robert W. Holley, Har Gobind Khorana and Marshall W. Nirenberg received the 1968 Nobel Prize in Medicine and Physiology for their interpretation of the genetic code and its function in protein synthesis.

The origin of cellular organelles

The standard genetic code diverges at organelles that are sites of the cell's energy production; these organelles are called mitochondria (singular: mitochondrion). By a divergence we mean that in some cases the codons will not coincide with the assignments shown in the standard code. At least three examples are known in which the mitochondrion genetic code diverges from the standard code: yeast, *Drosophila* and human mitochondria [11-13]. Symbiosis is a process by means of which individuals of different species interact. It has been assumed that the cellular organelles, such as the mitochondrion and the chloroplast, originated by means of an obligatory symbiosis, in analogy with what happens with the well known obligatory symbiosis between an alga and a fungus (the lichen) [14].

TABLE 4.3: The standard genetic code. The notation for the twenty amino acids follows the standard three-letter notation.

UUU	Phe	UCU	Ser	UAU	Tyr	UGU	Cys
UUC		UCC		UCC		UGC	
UUA	Leu	UCA	Ser	UAA	Stop	UGA	Stop
UUG		UCG		UAG	Stop	UGG	Trp
CUU	Leu	CCU	Pro	CAU	His	CGU	Arg
CUC		CCC		CAC		CGC	
CUA	Leu	CCA	Pro	CAA	Gln	CGA	Arg
CUG		CCG		CCG		CGG	
AUU	Ile	ACU	Thr	AAU	Asn	AGU	Ser
AUC		ACC		AAC		AGC	
AUA	Ile	ACA	Thr	AAA	Lys	AGA	Arg
AUG	Met	ACG		AAG		AGG	
GUU	Val	GCU	Ala	GAU	Asp	GGU	Gly
GUC		GCC		GAC		GGC	
GUA	Val	GCA	Ala	GAA	Glu	GGA	Gly
GUG		GCG		GAG		GGG	

One possible rationalization for the mitochondrion genetic code deviations is that symbiosis may have occurred very early in evolution, the possible scenario being that of a cellular predator invading a larger cell, such as *Thermoplasma* (Domain Archaea), thus presumably preceding the appearance of the eukaryotic cells themselves [15], most of which have adopted the standard genetic code. However, there remains the fact that some bacteria and protists do have the standard genetic code.

Hence, it is not evident how to establish a time chronology in the evolution of the genetic codes of the contemporary living cells.

From chemical to cellular evolution

We have only sketched two of the major steps (homochirality and the universal genetic code), which are significant in the pathway towards life during chemical evolution on

Earth. These steps should have taken place from 4.6 - 3.9   Gyr BP, the preliminary interval of geologic time which is known as the Hadean Subera [16].

It should be noticed that impacts by large asteroids onto the early Earth do not necessarily exclude the possibility that the period of chemical evolution may have been considerably shorter. Indeed, it should not be ruled out that the Earth may have been continuously habitable by non-photosynthetic ecosystems from a very remote date, possibly over 4 Gyr BP [17].

The content and the ratios of the two long-lived isotopes of reduced organic carbon in some of the earliest sediments (retrieved from the Isua peninsula, Greenland, some 3.8 Gyr BP), may convey a signal of biological carbon fixation [18].

This reinforces the expectation that chemical evolution may have occurred in a brief fraction of the Hadean Subera (4.6-3.9 Gyr BP, cf., Fig. 5.1), in spite of the considerable destructive potential of large asteroid impacts which took place during the same geologic interval in all the terrestrial planets, already referred to as the heavy bombardment period.

In subsequent suberas of the Archean (3.9 - 2.5 Gyr BP), life, as we know it, was present. This is well represented by fossils of the domain Bacteria, which is well documented by many species of cyanobacteria [19].

We may not exclude from the geochemical data earlier dates for the first prokaryotes. Indeed, the possible origin of life on Earth may have occurred immediately after the end of the Hadean Subera, some 3.9 Gyr BP.

Was there liquid water on the early Earth 4,400 million years ago?

A window to the earliest stages in the geologic evolution of our planet is possible form analysis of small grains of 'zircon'. These are minerals of great resistance to high pressure and temperature. For this reason they are indicators of the state of the early Earth, even though it was going through the 'heavy bombardment period' in which the collision of meteorites and larger bodies was more frequent than now. Specific analysis of the isotopes of elements (uranium and lead) present in a grain of zircon from the Narryer Gneiss Complex in Western Australia are consistent with the presence of continental crust and even liquid water between 4,400 and 4,300 million years before the present [20].

The oxygen isotopic composition of the zircon in question was analyzed. According to the standard method it should tell us something about the magma (molten rock) from which the zircons crystallized. By implication the nature of the rocks that gave rise to the magma can be inferred. It is known that heavy oxygen is produced by the interaction between rock and liquid water, a process that should occur at sufficiently low temperatures (to allow for the liquid state of water). The presence of the heavy isotope of oxygen in the zircons from the Narryer Gneiss Complex suggest that the magma had been produced by an episode of high temperature (or heavy pressure) on the surface of the early Earth, on which there was liquid water.

Given the significant implications for the first appearance of life on the surface of the early Earth these results will require further research. If such controls are done by independent teams, the first remarkable aspect for the science of astrobiology is that the mechanisms of chemical evolution and its transition to the first appearance of a single cell occurred in a temporal scale that might be much shorter than previously suspected, since the occurrence of liquid water could then be traced to just over 100 million years after the formation of the planet itself.

The reader should observe that organic material is another basic ingredient for the origin of life besides liquid water. As we mention in Chapter 3, a significant input was provided by incoming meteorites and comets. Finally, a third ingredient for the generation of life on Earth was the presence of sources of energy during the early geologic evolution of our planet. It has been evident since the early experiment of Stanley Miller that for prebiotic synthesis to take place, a possible source of energy must have been electric activity such as lightning (cf., Chapter 15: "Stanley Miller"); but another possibility, which shall be discussed later is, for instance, volcanic activity, either on the Earth's crust, or at hydrothermal vents.

The dawn of unicellular organisms

In spite of the fundamental work of Darwin that gave rise to modern biology, the fact remained that his contemporaries were still dividing the Earth biota into animals and plants.

It was only in the 1930s when taxonomy shifted its emphasis form the multicellular dominated classification to one more oriented towards basic cellular structure [21]. The encapsulation of chromosomes in nuclei was clearly absent in bacteria. This remark led to a division of all living organisms into two groups that went beyond the animal/plant dichotomy. As we have seen earlier, the new groups were firstly, the cells that lacked a nucleus (*karyon* in Greek); they were called appropriately 'prokaryotic'. Secondly, those cells that had nuclei, were called eukaryotic (i.e., truly nucleated). However, amongst bacteria rapid and random exchange of genes does occur. This is a phenomenon called horizontal gene transfer (HGT). This randomness deterred the use of sequences of the biomolecules for the construction of a tree of life (phylogenetic tree). Yet, Linus Pauling and Emil Zuckerkandl had pointed out that a molecular clock might be identified from the slow mutation rates of some biomolecules[22].

The question then is to identify the proper molecule that would not be affected by HGT, as evolution proceeds. Such molecules consist of RNA which make up ribosomes, together with some protein constituents. This form of RNA is called ribosomal RNA (rRNA).

Extensive work with these molecular 'chronometers' have led to a taxonomy of all life on Earth that is deeper and more significant than the dichotomy prokaryote/eukaryote. On the basis of evidence from molecular biology, a classification has been suggested in which the highest taxon is called a 'domain', instead of a 'kingdom'. In this approach, which was mentioned briefly earlier in this chapter (cf. "Extraordinary symmetry of cell membrane molecules"), there are three 'branches' in the tree of life:

• Archaea, whose most striking molecular characteristic is that the cellular membrane differs from the cellular membrane of the other two domains [24].

• Bacteria, encompassing all bacteria; their cellular membranes are more similar to those of the eukaryotes than the archaeans, and

• Eucarya, including all the truly nucleated cells.

This formulation of a proper taxonomy allows the discussion not only of the branches of the tree, but also of the root itself (the universal common ancestor) Alternatively the root of the tree has been called a 'progenote' or a 'cenancestor'. (The reader is advised to consult these terms both in the Glossary, as well as in the Subject Index). Since the formulation of Darwin's Theory of Common Descent, the nature of the universal common ancestor has remained an important problem in evolutionary biology.

The tree of life with a common root and trunk from which the three main branches developed early in the evolution of the Earth, are not perfect analogies. Recent molecular analysis suggests a somewhat more complicated picture [25]. Independently of the features of the phylogenetic tree, several aspects of the dawn of cellular life are emerging slowly.

An early mode of nutrition developed, in which microorganisms were able to use organic and inorganic substances as energy sources: these organisms are called chemotrophs. They are organisms that still live to this date.

Evidence is accumulating, which supports the possibility that organisms live and prosper deep in the terrestrial subsurface, at depths as great as several kilometers. Such environments may have been favored at the dawn of cellular life on Earth and elsewhere.

In the following chapter we shall learn that biogeochemical data suggests that the origin of life may have occurred even before the period of heavy bombardment was over. This means that practically before the time when the Solar System had completed the incorporation of planetesimals into the planets and small bodies that we observe today, the living process already may have been working its way into extreme niches, such as life underground.

Up to that time, soon after about 4 Gyr BP, planetesimals were still drifting into the path of the Earth orbit. These collisions are well documented, since the airless and geologically inert surface of the Moon testifies with its well-preserved record of craters, the colossal collisions that the surface of the Earth must have been suffering at that time. Thus microorganisms that could accommodate their life styles to extreme conditions would have been favored by the process of natural selection.

"Extremophile" is a term used sometimes for any of the many microorganisms that are capable of different degrees of adaptability to extreme ranges of living conditions. Many such 'extremophilic' microorganisms are known today. They have even been isolated from deep-sea hydrothermal vents [26].

Besides, today hydrogen sulfide and methane are abundant in the fluids of the deep-sea. From these unusual chemical sources chemothrophs obtain all the energy required for their survival. Such environments and microorganisms, as we have sketched in this section, are likely candidates for having given rise to the dawn of cellular life on Earth. These general considerations give some support to the thesis that similar ecosystems may emerge elsewhere as well.

BOOK 2:

EVOLUTION OF LIFE IN THE UNIVERSE

5

From the age of prokaryotes to the origin of eukaryotes

Our subsequent discussion is based on the idea that evolution has taken place on Earth. Darwin's major thesis was that evolutionary change is due to the production of variation in a population and the survival and reproductive success of some of these variants.

In this chapter we shall concentrate on the first stages of the story of life on Earth guided by Darwin's ideas. We pick up the story once the earliest and simplest living cell has already formed. As we saw in Chapter 4, it is sometimes called the progenote or, alternatively, the' cenancestor'.

A unique event lost somewhere deep in the Archean

Although we will be guided by Darwin, his ideas were developed in a different context from the origin of life. It was clear to Darwin that the tree of life had been evolving for a long period, incompatible with the estimates arising form a literal reading of the Bible. And like all trees it must have a root, which we have just called a cenancestor.

The evident unified nature of the tree of life speaks for itself, otherwise there would be growing a whole forest of life on Earth, a fact that contradicts everything that is known to naturalists. The origin of life for Darwin was a unique event lost somewhere deep in the Archean. The controversy that was raging in Victorian England between atheists and creationists was avoided altogether by Darwin, who kept origins out of his written work (cf., Chapter 15).

He was content with the thesis that the origin of life on Earth was inaccessible to 19th century science. As a naturalist Darwin was only concerned with life's subsequent development. The real question in his monumental *The Origin of Species* was whether the different species of the organisms he was very familiar with had a common ancestor. His expertise on a particular organism, the barnacle, was generally accepted. He had attempted to study the whole barnacle group, and to produce a definitive text, which included fossil barnacles as well.

As in all cases of good scientific approach to intricate questions, by narrowing down the problem, Darwin was able to document his case in favor of his two theories, natural selection as the principal force in evolution, and the *Theory of Common Descent* of all species on Earth. One particular question that this narrow approach to the problem allowed him was to keep away form ideological matters. We shall also attempt to avoid the ideological trap. It is nevertheless interesting to recall the essential arguments in direct opposition to Darwin, even by some of his distinguished friends.

We do this at this stage in anticipation of the philosophical aspects of astrobiology that we shall discuss in Chapter 14. Sir Charles Lyell (1797-1875) was the author of *Principles of Geology* (1830-1833). In his book he supported the view that the distant past is to be explained only by forces that we experience today (the doctrine known as uniformitarianism); it emphasized uniform processes of change in nature. What worried Lyell regarding the work of his friend Darwin was that Man would loose his special place in creation [1].

Lyell was protecting what in his view would be 'radical degradation'. Although Darwin's definite strategy in the narrow approach was to stick to species and stay away from the problem of creation, Lyell argued against Darwin's 'ugly facts'. Following Darwin's example, our limited objective will be to learn from the geological history of the Earth regarding the transition from the cenancestor to more complex cells. We do not attempt to follow up the detailed chemistry underlying the genesis of the cenancestor.

Role of oxygen and iron in eukaryogenesis

Several lines of research suggest the absence of current values of oxygen, O_2, for a major part of the history of the Earth. Some arguments militate in favor of Archean atmospheres with values of the partial pressure of atmospheric oxygen (O_2) about 10^{-12} of the present atmospheric level (PAL). We have already seen in the Introduction that the growth of atmospheric oxygen was due to the evolutionary success of cyanobacteria, which were able to extract the hydrogen they needed for their photosynthesis directly form water.

One of the chief indicators of the growth of atmospheric oxygen is shale, which is a rock that has played a role in our understanding of biological evolution. The onset of atmospheric oxygen is demonstrated by the presence in the geologic record of red shale, colored by ferric oxide. The age of such 'red beds' is estimated to be about 2 Gyr.

At that time oxygen levels may have reached 1-2% PAL, sufficient for the development of a moderate ozone (O_3) protection from ultraviolet (UV) radiation for microorganisms from the Proterozoic Era (cf., Fig. 5.1), which is the time ranging form 2.5 billion years before the present (Gyr BP) to 570 million before the present (Myr BP). In fact, UV radiation is able to split the O_2 molecule into the unstable O-atom, which, in turn, reacts with O_2 to produce O_3, which is known to be an efficient filter for the UV radiation.

The paleontological record suggests that the origin of eukaryotes occurred earlier than 1.5 Gyr BP. Some algae may even date from 2.1 Gyr BP [2], a period comparable to the first onset of red beds. This is still rather late, compared to the earliest available prokaryotic fossils of some 3.5 Gyr BP [3].

However, if we keep in mind certain affinities between eukaryotes and archaebacteria (such as homologous factors in protein synthesis), we may argue that archaebacteria and the stem group of eukaryotes may have diverged at about the same time [4]. This conjecture, combined with the lightest carbon isotope ratios from organic matter, implies that bacteria capable of oxidizing methane (CH_4) ('methylotrophs') may have been using methane produced by archaebacteria that were able to produce it as a by-product of their metabolism. From a careful analysis of such fossils, an age of 2.7 Gyr BP is assigned to eukaryogenesis.

Signs of evolution in Archean rock formations

There are Archean rock formations (which may be found up to 2 Gyr BP) that are significant in the evolution of life. These are compounds of dioxide of silicon (silica) and iron, which occur in layers. In reference to their stratified structure they are called "banded iron formations" (BIFs). The period in which the BIFs were laid out ended some 1.8 Gyr BP. In the anoxic atmosphere of the Archean, iron compounds could have been dispersed over the continental crust.

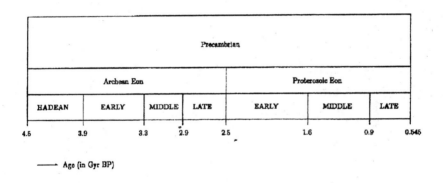

Figure 5.1: Stratigraphic classification. The Precambrian.

They could have absorbed some oxygen, thereby protecting photosynthesizers that could not tolerate oxygen. Such microorganisms in turn produced oxygen that combined with their environment to produce iron oxide (for example, hematite Fe_2O_3), which makes up the BIFs.

In strata dating prior to 2.3 Gyr BP it has been observed that there is an abundance of the easily oxidized mineral form of uranium (IV) oxide (urininite, for example the well-known variety *pitchblende)*. This argument supports the conclusion that we had to wait until about 2 Gyr BP for a substantial presence of free O_2. Once the eukaryotes enter the fossil record, their organization into multicellular organisms followed in a relatively short period (in a geological time scale).

First appearance of multicellular organisms

Metazoans are assumed to have arisen as part of a major eukaryotic radiation in the Riphean Period of the Late Proterozoic, some 800-1,000 Myr BP [4]. when the level of atmospheric O_2 had reached 4-8% PAL (cf., Fig. 5.1). There is some evidence in the Late Proterozoic during the Vendian Period, for the existence of early diploblastic grades (Ediacaran faunas).

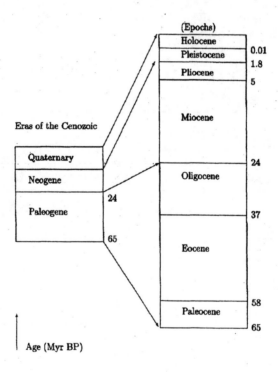

Figure 5.2: Epochs of the Cenozoic Era.

These organisms were early metazoans with two germ layers, such as the modern coelenterates (jellyfishes, corals, and sea anemones). Later on, when the level of atmospheric oxygen had reached values in excess of 10% PAL, these groups of primitive animals were overtaken in numbers by more complex groups, triploblastic phyla (cf., Glossary), as the level of atmospheric oxygen had reached 40% PAL.

These organisms are called the Cambrian faunas, i.e., Early Paleozoic faunas, some 500 million years before the present (Myr BP), which were mainly metazoans with three germ layers. They constitute at present the greater majority of multicellular organisms, including *H. sapiens sapiens.*

We may obtain further insights from paleontology: acceleration in the evolutionary tempo is observed after the onset of eukaryogenesis, as it is clearly demonstrated by the microfossils of algae from the Late Proterozoic [5] and by the macrofossils of the Early Phanerozoic [6]. Such evolutionary changes within the first billion years of atmospheric oxygen raised the simple prokaryotes to eukaryotes, animals and plants.

The origin of chromosomes

The most remarkable period subsequent to the establishment of the genetic code, or more precisely to the establishment of the translation mechanism of the genetic message, concerns the period in which the chromosomes originated at the very beginning of the evolution of cells. It is widely believed that the intron-exon structure of genes was part of the first cell. This view implies that the genes of the most advanced higher organisms may reflect the primitive genome structure more clearly than the genes of prokaryotes, as during their evolution bacteria and archaebacteria generally have eliminated their 'introns' [7].

This idea is persuasive enough for looking at the contemporary eukaryotic cell as a potential source of information on the structure and function of the first cell. This viewpoint allows the possibility of facing even some of the most difficult questions concerning the primordial cell.

Perhaps amongst such difficult questions stands out the origin and evolution of chromosomes. One general guiding line in this important aspect of the origin of the progenote [8] is to explain the past in terms of processes that may currently be present in cells. It seems likely that changes in a protocell that led to the eukaryotic cell were in some way related to the control of gene expression [9].

Unfortunately, there is yet no general understanding of this aspect of the eukaryotic cell. One way of approaching this intricate problem is to extract from the known phenomena of genetics some general aspects of the relations between the main relevant parameters and to use such knowledge to sketch possible aspects of the earliest chromosomes.

A very tentative start in this program has already been published [10-11]. The folding problem of the macromolecules of life seems an appropriate starting point in our search for insights of the earliest aspects of the cell. Some insights have been gained in understanding the underlying mechanism that controls the folding of any polypeptide chain into unique three-dimensional protein structures.

On DNA folding

We may consider the problem analogous to protein folding, but instead with the DNA that makes up the bulk of chromosomes. DNA folding is characterized by the '100-Å nucleosome filament' [12]. This is of fundamental importance to the question of gene expression. In turn, the more complex process of gene expression in the nucleated cell may be related, as we stated above, to the transition into eukaryogenesis.

There is a hierarchy of levels of folding beyond the 100-Å nucleosome filament: The next level of complexity is provided by 300-Å filament, which is arranged into a solenoidal configuration with about six nucleosomes per turn. During interphase in the cell cycle it is this solenoidal arrangement that constitutes the most abundant form of chromatin.

However, at later stages in the cell cycle this structure serves as the basis for further folding, ending up at the highest degree of folding observed at the metaphase chromosome. In view of the basic role played by all the stages of the hierarchy, it is of certain interest to find a formalism by means of which we may anticipate the regular manner of DNA folding. This problem is of considerable difficulty.

We have thus restricted our attention to DNA folding in the later stages of chromatin compaction. The biochemical basis for the difference between heterochromatin (the more compact structure of chromatin [13]) and euchromatin (its less compact form), remains unknown. In both cases the 100-Å nucleosome filament contains approximately the same ratio of DNA to protein. Heterochromatin appears most frequently at the center and ends of the chromosomes [14]. Some models attempt to account for movement of the group of proteins that are involved in the replication of DNA [15].

These preliminary attempts at modeling the replication process in eukaryotes may be seen as preliminary efforts to understand invariant aspects of genetics that could give some insights into the early stages of eukaryogenesis. But much progress remains to be done before these insights give us a complete understanding of the processes that led to eukaryogenesis on Earth.

The Molecular Clock hypothesis

Molecular changes during evolution may be used in order to reconstruct phylogeny. This possibility, mentioned in Chapter 4, is of considerable importance in phylogeny; it is referred to as the 'molecular-clock' hypothesis [16].

The molecular changes of sequences of nucleic acids that take part in the universal functions of the cell have proved to be particularly fruitful. These studies are based mainly on some of the ribosome molecules. (These studies are based on the RNA molecule of one of the two subunits that make up the ribosome; specifically, the small one.).

These ribosome molecules have a key to phylogeny of the entire span of living cells. Such RNA molecules also allow a natural classification of single-cell organisms.
This approach may be particularly useful in cases where morphology may be insufficient. A clear example is the difficult phylogenetic position of onychophorans (velvet worms). These animals share features of annelids (segmented worms) and arthropods (jointed-foot invertebrates); they resemble mollusks and, like mammals, have placentas.

On morphological grounds it has been assumed that these animals belong to a separate phylum, but recent evidence from certain ribosomal RNA sequences suggests rather that onychophorans belong to the phylum Arthropoda [17]. The fact that the eukaryotic cell is a multiple-genome organism has been brought up in discussions of the molecular-clock hypothesis (in which one gene is taken to represent a whole organism).

However, the concept of phylogeny is meaningful as long as a clear majority of the essential genes in a genome share a common heritage [18]. Nevertheless, some care may be needed [19], in view of the proposal for introducing the domains Archaea, Bacteria and Eucarya [20], beyond the well-established five-Kingdom classification [21].

It is possible that taxa above the kingdom level may obscure the fundamental division of the biota into organisms with a single ancestry and those with multiple genomes evolving by symbiosis. Besides, the question remains of whether certain sets of molecular sequences should take precedence over other sets. This point may be underlined by the fact that plant cells have normally three genomes, while certain algae have four genomes; for example, the cryptomonads. In this case the fourth genome is in an organelle (the nucleomorph) bounded by a double membrane, possibly the remains of the nucleus of a symbiotic eukaryote [22]. Such cases are referred to as 'secondary symbioses'.

A transition in the pathway towards intelligence

Having discussed the origin of chromosomes, one of the questions we must proceed to face is eukaryogenesis. This is perhaps the most significant event in the diversification of life and, according to the fossil evidence, it may have occurred not later than in the Proterozoic about 1.8 - 2.0 Gyr BP. Eukaryotes have their DNA linked in chromatin; the main organelles, i.e., mitochondria and chloroplasts, are normally in their cytoplasm.

However, protozoans may provide examples of mitochondrion-less eukaryotes. There are even phyla of amitochondrial protozoa; for instance, the microsporidia [23]. Besides, in two out of three Eucarya kingdoms of multicellular organisms (i.e., Animalia and Fungi) chloroplasts are absent.

The origin of these two types of organelles, mitochondria and chloroplasts, in the eukaryotes is to be found, according to the serial endosymbiosis hypothesis, in separately evolved organisms.

Thus, at least the origin of the red alga chloroplast may be traced back to cyanobacteria (this may be illustrated with the analysis of ribosomal RNA of the unicellular marine red alga *Porphyridium* [24]).

On the other hand, mitochondria may be linked with purple bacteria resembling *Paracoccus denitrificans*. This prokaryote is suggested to be a plausible ancestor, because when all various biochemical parameters are taken into account, *P. denitrificans* resembles a mitochondrion much more closely than other aerobic bacteria [25].

In fact, symbiosis implies that the ancestral prokaryote may have been taken up by a chloroplast-free amoeboid protoeukaryote; eventually the symbiont may have lost autonomy, possibly by horizontal gene transfer between the protomitochondrion and the host's nucleus.

These events, which are different from natural selection, were factors that would have led to the evolution of a single-cell organism that had already integrated the metabolism of the partners in symbiosis. Further support for symbiosis is that both eukaryotic organelles we have just mentioned, have their own separate mechanisms of translation of the genetic message:

In Chapter 4 we have seen that separate genetic codes are known for each organelle (cf., "Origin of cellular organelles", p. 62). The origin of the nucleus does not seem to be explained by symbiosis. The main reason for sustaining the thesis of 'direct filiation' (i.e., differentiation) has been summarized briefly [26]. This sets some limitation on the extent to which symbiosis may have shaped the first eukaryotic cell, leaving the question of the nature of the earliest eukaryote as an open problem.

Primitive eukaryotes have been studied in sufficient detail to allow the study of their origins. One such taxon is the family Cyanidiophyceae (this group of organisms are rhodophytes, commonly known as red algae) [27].

In which environments can extremophiles survive?

It is useful to question in which atmospheres extremophiles may survive, as we now know of several solar-system planets and satellites that have atmospheres, not all similar to our own. For example, four large planetary satellites in the outer solar system are known to have atmospheres [28]: Europa and Io (Jupiter), Titan (Saturn) and Triton (Neptune).

Mars currently has a CO_2 dominated atmosphere. Besides, Ganymede (a Jovian satellite) is known to have a tenuous oxygen atmosphere [29]. Two satellites of Saturn, Rhea and Dione, which orbit within its magnetosphere have accumulated oxygen and ozone, since their icy surfaces have been exposed to ion irradiation [30]. The possibility of extending the biosphere deep into the silicate crust of Mars has some implications. This question seems pertinent to astrobiology.

We cannot exclude the possibility that organisms, which have been found to inhabit deep in the silicate crust of the Earth, may been deposited with the original sediment, and survived over geologic time. This question has been considered in some detail from the point of view of geophysics [31].

Mechanisms of evolution beyond natural selection

Before we consider the evolution of man it is worth delaying the discussion in order to appreciate how eukaryogenesis took place on Earth.

We have learnt in the previous chapter ("The dawn of unicellular organisms", p. 65) that an important genetic mechanism beyond natural selection has been called: horizontal gene transfer (HGT). HGT is a process by means of which genetic information may be implanted into [32-34]:

• a target species from a donor species, or

• intracellularly between organelles, or

• between organelles and the cell nucleus. Some genes may have been exchanged between the chloroplast and the nucleus.

• The complete sequence of the chloroplast genome of a bryophyte [*Marchantia polymorpha*, a liverwort] has been established. Its DNA has some genes that are not detected in the chloroplast of a more complex plant species [tobacco].

On the other hand, some genes may then have been exchanged between the nucleus of the host plant cell and the residual genome of the corresponding chloroplast during the evolution of angiosperms. Such HGTs may have occurred during the time span separating the first appearance of bryophytes (Paleozoic), from the first appearance of the angiosperms in the late Mesozoic.

Constraints on chance

It is instructive to appreciate that there are several constraints on chance. Some examples are relevant to the question of whether life elsewhere might follow pathways analogous to the transition from prokaryotes to eukaryotes as known to us from the only example that we know, namely the Earth biota. From various examples of constraints on chance enumerated elsewhere [35], we select three and provide an additional example:

1. Not all genes are equally significant targets for evolution. The genes involved in significant evolutionary steps are few in number, they are the so called regulatory genes. In these cases mutations may lead to unviable organisms and are therefore not fixed.

2. Once a given evolutionary change has been retained by natural selection, future changes are severely constrained; for example, once a multicellular body plan has been introduced, future changes are not totally random, as the viability of the organisms narrows down the possibilities. For instance, once the body plan of mammals has been adopted, mutations such as those that are observed in *Drosophila*, which exchange major parts of their body, are excluded. Such fruit-fly mutations are impossible in the more advanced mammalian body plan.

3. Not every genetic change retained by natural selection is equally decisive. They may lead more to increasing biodiversity, rather than contributing to a significant change in the course of evolution. This may be illustrated with the following example: Within the Solanaceae family one tomato chromosome has a region between its center and end ('centromere' and 'telomere'), which consists of a row of segments in which DNA is compacted into tight masses, largely inactive in transcription ('chromomeres'[36]) in *Petunia*, in spite of being another genus of the same family, the abundance of chromomeres is not preserved, since larger blocks of heterochromatin are observed. A tiny bit of heterochromatin may be superficially indistinguishable from a eukaryotic 'chromomere'. These two genera illustrate how quickly evolution can induce rearrangements of heterochromatin, while preserving general chromosome structure; this mutation has not contributed any significant change in the course of evolution.

Chance, contingency and convergence

Besides the constraints on chance mentioned in the last section, we should recall the eternal confrontation deep in the fabric of evolutionary theory, brought to popular attention by Jacques Monod in *Chance and Necessity*.

Indeed, implicit in Darwin's work we have *chance* represented by the randomness of mutations in the genetic patrimony, and their *necessary* filtering by natural selection. However, the novel point of view that astrobiology forces upon us accepts that randomness is built into the fabric of the living process. Yet, contingency[37], represented by the large number of possibilities for evolutionary pathways, is limited by a series of constraints as mentioned before. What we feel is even more significant for astrobiology, is to recognize that natural selection necessarily seeks solutions for the adaptation of evolving organisms to a relatively limited number of possible environments. We have seen in cosmochemistry that the elements used by the macromolecules of life are ubiquitous in the cosmos. We have also seen that the formation of solar systems is limited by a set of physical phenomena, which repeat themselves for instance, in the Orion Nebula [38]; those jovian planets discovered in our 'cosmic village' (cf., Chapter 10) can bear satellites in which conditions similar to those in the satellite Europa may be replicated (cf., Chapter 8).

To sum up, the finite number of environments forces upon natural selection a limited number of options for the evolution of organisms. From these remarks we expect convergent evolution to occur repeatedly, wherever life arises. It will make sense, therefore, to search for the analogues of the attributes that we have learnt to recognize in our own particular planet.

We postpone the conclusion of the present discussion till the end of Chapter 11, when we shall discuss the central question of astrobiology, namely, is life necessarily linked to the Earth?, or alternatively, did the origin of life on Earth produce only an infinitesimal fraction of life in the cosmos?

6

Eukaryogenesis
and evolution of intelligent
behavior

In evolutionary terms, we choose not to emphasize complex multicellular organisms. Instead, we have shifted our attention to the single-celled nucleated organism (eukaryote), whose evolution is known to have led to the evolution of intelligent behavior, at least on planet Earth.

Modern taxonomy emphasizes single-celled organisms

This point of view is forced upon us by the present taxonomic classification of organisms into domains, which stresses single-celled organisms. Previously, an older taxonomic classification, in terms of kingdoms as the highest taxa, highlighted multicellular organisms, incorrectly in our view.

The older approach was due to biologists lacking an understanding of molecular biology which, as we stated in the Introduction, only made its first appearance in the early 1950s. Biology has been able to provide us with sufficient insights into the cell constituents to permit the wide acceptance of a comprehensive taxonomy.

This new approach places Bacteria, Archaea and Eucarya as the highest groups (taxa) of organisms. We can paraphrase Sir Julian Huxley's comments in the introduction of *"The Phenomenon of Man"* [1] by remarking that there is an evident inexorable increase towards greater complexity in the transition from Bacteria to Eucarya.

To borrow the phrase of Christian De Duve [2], we may say that the laws of physics and chemistry imply an 'imperative' appearance of life during cosmological evolution, a view which is not in contradiction with the relevant critical remarks of Sir Peter Medawar in his comment on Père Teilhard's work [3].

The phenomenon of the eukaryotic cell

We may argue that not only life is a natural consequence of the laws of physics and chemistry, but once the living process has started, then the cellular plans, or blueprints, are also of universal validity: the lowest cellular blueprint (prokaryotic) will lead to the more complex cellular blueprint (eukaryotic). This is a testable hypothesis.

Within a decade or two, a new generation of space missions may be operational. Some are currently in their planning stages, which are aiming to reach the Jovian satellite Europa in the second decade of this century [4]. We shall present the rationalization behind this effort in the present chapter, and discuss some details in Chapter 8.

Closely related to the above hypothesis (the proposed universality of eukaryogenesis), concerns the different positions which are possible regarding the question of extraterrestrial life:

Is it reasonable to search for Earth-like organisms, such as a eukaryote, or should we be looking for something totally different?

We will discuss in turn some of the arguments involved. Firstly, the more widely accepted belief on the nature of the origin of life is that life evolved according to the principles of deterministic chaos [5]. Evolutionary developments of this type never run again through the same path of events. Secondly, the possibility for similar evolutionary pathways on different planets of the solar system has been defended recently [6].

Indeed, even if some authors may consider this to be a remote possibility, there is an increasing acceptance that catastrophic impacts may have played an important role in shaping the history of terrestrial life.

There may be some common evolutionary pathways between the microorganisms on Earth and those that may have developed on Mars during its 'clement period' (roughly equivalent to the early Archean in Earth stratigraphy). The means of transport may have been the displacement of substantial quantities of planetary surface, due to large asteroid impacts on Mars.

Finally, even in spite of the second possibility raised above, many researchers still see no reason to assume that the development of extraterrestrial life forms followed the same evolutionary pathway to eukaryotic cells, as it is known to have occurred on Earth. Moreover, it would seem reasonable to assume that our ignorance concerning the origin of terrestrial life does not justify the assumption that any extraterrestrial life form has to be based on just the same genetic principles that are known to us.

In sharp contrast to the position denying that common genetic principles may underlie the outcome of the origin of life elsewhere, there is a fourth way of approaching the question of the nature of extraterrestrial life. We may conclude that we all agree that the final outcome of life evolving in a different environment would not be the same as the Earth biota.

New ground is reached raising the question: *How different would be the outcome of the origin of life elsewhere?* [7] This has led to a clear distinction that there is no reason for the details of our phylogenetic tree to be reproduced elsewhere (except for the possibility of biogenic exchange in the solar system discussed above).

The evolutionary tree of life constituted by the Earth biota may be unique to planet Earth. On the other hand, there is plenty of room for the development of differently shaped evolutionary trees in an extraterrestrial environment, where life may have taken hold.

The phenomenon of multicellularity

A point worth emphasizing is that higher organisms are not only characterized by eukaryoticity, but also by multicellularity. This second feature evolved gradually from unicellular microorganisms. An advantage of cellular aggregation is its adaptive value. An evident example of such advantage regards predator-prey interactions [8]. This follows from direct observation of contemporary colonies of the bacterium *Myxococcus xanthus*. This bacterium forms spherical colonies which increase their success in predation. The unicellular aggregation lets the prey enter the sphere through gaps, which are able to retain digestive enzymes that are produced by the bacteria. We have seen in previous chapters that the prokaryotic blueprint is a consequence of chemical evolution. (In a geologic time scale prokaryogenesis is almost instantaneous.) It follows from these considerations that multicellularity is also a consequence of chemical and biological evolution in a terrestrial-like environment.

The detailed current view on terrestrial evolution spanning from the first appearance of the eukaryotic cell to the evolution of multicellular organisms is as follows: In the Proterozoic Eon, a geologic period which extends from 2.5 to 0.545 Gyr BP, we come to the end of the Proterozoic, an era which had seen the first appearance of the eukaryotic cell. The Proterozoic was followed by the most recent eon, that has seen the spread of multicellular organisms throughout the Earth on its oceans and continents, the Phanerozoic, which extends from the end of the Proterozoic to the present. We should comment on some aspects of the current Phanerozoic:

• The Paleozoic Era saw the first appearance of fish (510 Myr BP) and some other vertebrates, a landmark in the evolution of life on Earth.

• The Mesozoic Era, initiated after the massive extinction some 250 Myr BP. This era saw the first appearance of mammals (205 Myr BP).

Current evolutionary thinking sees the first appearance of humans, a question of central interest both to theology (cf., Chapter 13) and philosophy (cf., Chapter 14), as a natural consequence of the series of events which began in the Mesozoic and culminated in the Cenozoic Era, particularly some 5 Myr BP when, according to anthropological research, our human ancestors first appeared.

Evolution of the hominoids

The last era of the Phanerozoic Eon is called the Cenozoic (cf., Fig. 6.1). This was initiated after the great extinction of the Upper Mesozoic . The Cenozoic saw the first appearance of primates some 60 Myr BP. These essentially small animals took up residence in trees. Natural selection eventually provided these small mammals with long arms, fingers and frontal vision. Some of these species remained unaltered to the present. They are represented by the lemurs of Madagascar, which include the indri, a prosimian that hardly ever comes to the ground [9].

The Rift Valley in Eastern Africa may hold the key to the divergence of hominids from the great apes (represented by chimpanzees, gorillas and orangutans). In fact, the Rift Valley was created by tectonic forces about 8 Myr BP. The new mountain boundaries divided the hominoid primates, which are a group that includes the lesser apes (gibbons, the small arboreal anthropoid apes of south-eastern Asia), the great apes, and humans.

Such a division formed two groups. Firstly, a western group bound to the forests, which were the ancestors of the modern great African apes; secondly, an eastern group was forced to live in the savanna, evolving directly into our ancestors, the first humans. Intelligence, in its highest form, had to await the emergence of the genus *Homo*, some 2 Myr BP, when its first traces arose in our ancestors. A more evident demonstration of the appearance of intelligent behavior on Earth than the habilines' tools (cf., Chapter 12), or the ceremonial burials had to wait till the Magdalenian 'culture' (cf., Table 6.1). This group of human beings flourished from about 20,000 to 11,000 years BP.

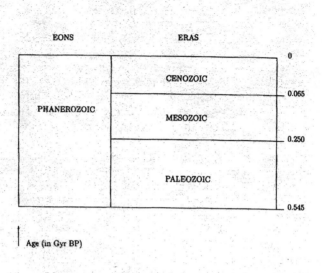

Figure 6.1. Eras and suberas of the Phanerozoic Eon, the most recent eon of Earth history.

The Magdalenians left some fine works of primitive art as, for instance, the 20,000 year old paintings on the walls of the cave discovered in December 1994 by Jean-Marie Chauvet in south-eastern France. Indeed, the birth of art in the Magdalenians' caves is one of the most striking additions to the output of humans that entitle us to refer to the groups that produced these fine works, as *cultures,* rather than *industries,* a term which is reserved to the group of humans that produced characteristic tools, rather than works of art.

The purpose of this sketch has been to bring into focus different branches of science. These include the social sciences. In terms of them some answers have been provided to the questions humans have recorded since the most ancient times. Pioneers in these queries were the Israelites over two thousand years ago, when the Old Testament of the Bible was written down. Those disciplines include anthropology, paleontology, prehistoric research, geochronology, biogeology and geochemistry, amongst others.

Evolution of intelligent behavior in the hominids

Amongst the earliest humans, or hominids, the habilines (*Homo habilis)* were a group from East Africa that first appeared from about 2.0 to 1.5 Myr BP, just before the onset of the Quaternary Period about 1.8 Myr BP.

The brain of these early hominids was large, 800 cc, but still about 60% of the size of our own brain (1360 cc, on the average, over the last 100,000 years). In spite of this difference in cranial capacity, there is evidence that the habilines used a wide variety of tools, demonstrating that they were more intelligent than the australopithecines.

The group of tools produced by the habilines belongs to the so called "Oldowan industry", as they have been found in the Olduwai Gorge, where the East African Rift cuts through about 100 meters of sediments laid down in a former lake basin.

More advanced tools are those of the Acheulian industry, which gradually replaced the earlier tools, but the new tradition in tool-making is characteristic of a more advanced hominid, *Homo erectus,* who lived from at least 1.7 to 0.5 Myr BP, and may have been the first hominids to use fire for cooking. The Acheulian industry, in turn, was gradually replaced by the Mousterian, which emerged during the last interglacial and lasted some 30,000 years.

Glaciations and interglaciations were major challenges to the development of the genus *Homo*, from the habilines to modern humans. The most recent hominids besides ourselves are the Neanderthals (*Homo sapiens neanderthalens*), who lived in Europe and western Asia, approximately from the Riss glaciation (250,000 - 120,000 yr BP) to the Würm glaciation (80,000 - 10,000 yr BP), most of them living from 100,000 to 40,000 yr BP (cf., Table 6.1). Instead of considering the concept of intelligence itself, the evolutionary background of this work induces us to stress cultural evolution with results that became more prominent since the last glaciation (the Holocene Epoch). These included the ability of the early humans to learn, communicate and teach their offspring skills, such as writing (some 5,000 yr BP), and even abstract and practical concepts such as religion, philosophy and rudimentary science.

We have seen in Chapter 1 that these aspects of humanity were already established some 2,500 yr BP. Such abilities of *H. sapiens sapiens* gave, to those that possessed them, increased advantage over the already selective advantage they had on other primates, which had been provided by genetics.

Cultural evolution may be considered as a consequence of the presence of larger brains. Not only humans, but even other mammals, and birds, possess an additional selective advantage proportional to brain size. This advantage goes beyond favorable random mutations, as the capacity to learn is an additional pressure that will encourage, not only the adaptability to the environment, but also some special attributes, such as learning.

Does eukaryogenesis lie in the pathway towards universal communication?

We would like to stress that cultural evolution is a stage in human development that can be traced back to that crucial transition from prokaryotes to eukaryotes, which is estimated to have occurred on Earth some 2 Gyr BP. We are now aware of the existence of extrasolar planets. This suggests the question whether in those environments eukaryogenesis may have already occurred. Data available from a large number of earth and life science disciplines, mentioned above, lead us to the conjecture:

Provided the planets of a given solar system have the appropriate volatiles (particularly water and oxygen), not only prokaryotic life will appear, but eukaryogenesis will take place.

This conjecture has the merit that it is subject to experimental verification, as we shall see in Chapter 8. The conjecture is not idle for it addresses a latent problem that underlies a large research effort, namely that of the search for extraterrestrial intelligent behavior (cf., Chapter 11). In fact, a defining attribute of humans is our cognitive ability. We seem to be the only species that has evolved on Earth due to evolutionary pressures in a terrestrial environment, which is capable of wondering, as we are doing in this book, about the relation life-universe.

TABLE 6.1: Archaeological classification with a reference to the European cultures. Some details are given of events, which have been relevant to the genus *Homo*. The continental classification for glaciations and interglaciations is given [10, 11].

Time $(10^3$ yrs)	Stratigraphical reference	Glaciations	Archaeologic Classification (General)	Archaeologic Classification (SW Europe & North Africa)
0			Iron Age	
			Bronze Age	
5	Holocene	Post Glacial	Neolithic	
			Mesolithic	
10				
10			P	
20			A	Magdalenian
30			L	
40	Late		E	
50	Pleistocene	Würm	O	
60				Mousterian
70				
80				
120		Interglacial	L	
250		Riss	I	
300		Interglacial	T	
500	Middle Pleistocene	Mindel	H	Acheulian
800		Interglacial		
900	Early	Günz	I	
1400	Pleistocene	Interglacial	C	
1700	Proterozoic	Danube		

However, we are not the only species where some degree of intelligent behavior has evolved. (In this context we define intelligence in the usual manner, in other words, the faculty of understanding, the capacity to know or apprehend.) A lesser degree of intelligence in this sense is also present in dolphins and other cetaceans in the aquatic environment (cf., Chapter 12).

The problem of communication amongst dolphins and between dolphins and humans [12] can be seen as a model study of a larger issue that we will have eventually to face, namely communication between our civilization and those potential products of natural selection on extrasolar planets, or their satellites. We shall return to the significant analogies between delphinids and anthropoids in Chapter 12.

In the wider scenario of the present work, the problem of communication amongst intelligent beings from different stellar environments has eventually to be faced in truly scientific terms (cf., Chapter 11). We consider the two evolutionary pressures, terrestrial and aquatic, that have given rise to the largely different brains of dolphins and humans (although some general features are common to both of them, such as the modular arrangement of neurons [13]).

The intelligent behavior of humans and dolphins are products of an evolutionary pathway that began from the general prokaryotic blueprint appearing on Earth almost instantaneously (in a geologic time scale, cf., Chapter 5), and continued through the sequence of eukaryogenesis, neuron, multicellularity and, finally, brains. The present discussion of universal aspects of communication amongst humans at some point has to be extended to all life in the universe. In other words, in spite of the fact that ours is the first species we are aware of that has evolved cognitive ability, our discussions should not be simply confined to our phylogenetic tree.

The origin of language in the universe

This question is of fundamental importance to astrobiology. In particular, it is relevant to the search of other civilizations. Language is a natural target for astrobiological research. It is a difficult subject, one of the reasons being that brains do not fossilize. The origin of language is not only a difficult problem from the point of view of embryological development (the human brain has over 10^{13} neurons), but also the historical perspective is difficult to assess. In our planet natural selection seems to have produced universal characteristics in the first species that has reached cognitive ability. Indeed, Steven Pinker in *The Language Instinct* argues [14] that language is compatible with gradual evolution due to natural selection. Pinker argues his case with a large amount of evidence in favor of a genetic basis for spoken language.

This possibility should be seen against the background of the development of a science of language. Noam Chomsky argued that language has underlying structural similarities. Since the evolutionary process has not led to our species having the faculty of expressing itself in one particular language, Chomsky defends the thesis that powerful constraints must be operative restricting the variety of languages [15]. Hence, according to this view, one can conclude that we must be born with knowledge of an innate Universal Grammar (UG). He supplied linguistics with a philosophical foundation (rationalism rather than empiricism).

According to Chomsky the structure of language is fixed in the form of innately specified rules: during the process of development all a child has to do is to turn on a few 'biological switches' to become a fluent speaker of a given language. In his view, children are not learning at all, anymore than birds learn their feathers[16]. Yet this point of view has turned out to be controversial.

Pinker argues convincingly in favor of the adaptationist interpretation of the origin of language. In this view, language can be seen as the 'imperative' by-product of *natural selection*, rather than a UG. The outline of his argument is as follows: the abrupt increase in brain size may have been due to the evolution of language, instead of language arising only as brain size goes beyond a certain threshold.

This faculty of the brain is to be searched as a consequence of natural selection, as any other attribute that characterizes a human being, such as size or skin color. Language, in this evolutionary context, evolved through social interaction in our hominid ancestors, as they adopted collective hunting habits. Better communication by means of rudimentary language would have been favored by natural selection.

Underlying all our efforts in astrobiology, we find the assumption that the laws of science are universal. Independent of the particular form given to an alien being by the evolutionary process on an extrasolar planet, or satellite (cf., Chapter 10), it seems reasonable to conjecture that Darwin's seminal theory of evolution through natural selection is probably the only theory that can adequately account for the phenomenon that we associate with life. This view has been defended by some biologists for almost 20 years, beginning with Richard Dawkins [17].

The co-evolution of brain and language

According to the ideas expressed in Chapter 12, brain is bound to evolve, once the evolutionary process is set in motion anywhere in the cosmos. Provided there is sufficient geologic time available to a given solar system (cf., comments in Chapter 1 regarding the differential ages of stars that follow from the information summarized in the HR diagram), natural selection is bound to play analogous roles *on brain evolution across species,* within a given phylogenetic tree.

Yet, we know from life on Earth that there are clear examples of evolution prizing one specialized faculty; this remark can be illustrated with the enormous development of the corresponding part of the brain. For example, 7/8ths of a bat's brain is devoted to hearing [18]. The thesis of co-evolution of language and brain has already been developed in considerable detail [12].

Without entering into too much detail, it is sufficient to point out that recent human evolution demonstrates that certain adaptations useful to other animals (for instance, fur and early sexual maturity) were, in fact, lost in our own species in favor of language, speech and consciousness. The presence of such faculties in humans that have evolved on our own solar system approximately half way through the lifetime of its star (i.e., the Sun), would argue in favor of *evolution of intelligent behavior in the plurality of new worlds.* We anticipate to the reader that the possibility of a large number of possible extra-solar worlds was brought to our attention in the last few years of last century (cf., Chapter 10).

BOOK 3:

DISTRIBUTION AND DESTINY OF LIFE IN THE UNIVERSE

On page 88

UPPER IMAGE (a)

The distribution of life in the universe will begin to be understood with the exploration of possible environments inside our own solar system where life can evolve. These artist's drawings depict a proposed model of the subsurface structure of the Jovian moon, Europa. Geologic features on the surface, imaged by the Galileo spacecraft might be explained by the existence of a layer of liquid water with a possible depth of more than 100 kilometers. (Courtesy of NASA.)

LOWER IMAGE (b)

An artist's sketch of the possible distribution of liquid water on Mars. This planet is currently one of the two most attractive possibilities for detecting the presence of life elsewhere in our solar system, the other possibility is Europa, c.f., previous illustration. (Image courtesy of the European Space Agency (ESA), (c) 2001.)

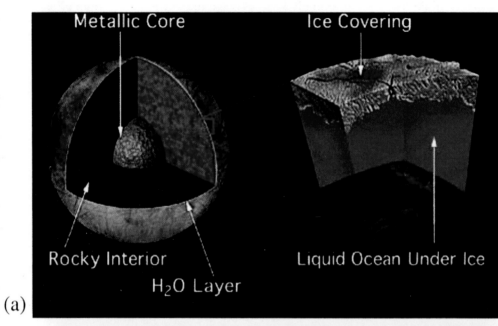

(a)

(b)

Part I

EXOBIOLOGY OF THE SOLAR SYSTEM (*):
SCIENTIFIC BASES FOR THE STUDY OF THE LIFE OF
OTHER WORLDS

(*) Before proceeding to Book 3, the reader is advised to consult the following terms
in the Glossary: astrobiology, bioastronomy and exobiology.

A first step towards a better understanding of the distribution of life in the universe is to investigate places in our own solar system where life may be present. This is just one of Mars Express mission's main objectives. This will be a European orbiter and lander mission. (Image courtesy of the European Space Agency (ESA), (c) 2001.)

7

On the possibility of biological evolution on Mars

The main interest in Mars from the point of view of astrobiology is centered on the fact that in the past this planet may have been hospitable to life. The details of the search for life on Mars shall be discussed briefly in this chapter.

Could there have been chemical evolution on Mars?

In Table 7.1 we begin with a review of the statistics of Mars. A preliminary question in the problem of life in Mars is whether there could have been chemical evolution there. There is a long history of experimental efforts with inorganic substances which have the ability to rotate the plane of polarization of light passing through a crystal. This ability is due to their chiral crystal structure. In this respect quartz, the most common form of crystalline silica (SiO_2), is found in nature in well defined enantiomorphic crystals in which the handedness is caused by the spiral atomic structure.

Quartz, therefore, is an obvious candidate for the stereoselective adsorption of a given enantiomer from a racemic mixture. A long series of experiments has led to the conclusion that asymmetric adsorption on quartz is a well authenticated phenomenon. It is also equally clear that in this process quartz acts merely as a passive 'filter' of homochiral domains, passive in the sense that quartz itself does not induce any catalytic process that may convert racemic or achiral reactants into optically active products [1]. Asymmetric adsorption of amino acids at the surface of chiral crystals is a possible physical mechanism that may have led to the onset of homochirality in one class of biomolecules. It is well documented that about one per cent more left-handed amino acids than right-handed amino acids are adsorbed on left-handed quartz crystals.

TABLE 7.1: Physical parameters of Mars.

Parameters	Values
Mass (Earth mass=1)	0.1
Radius (Earth radius =1)	0.53
Orbital eccentricity	0.0934
Eccentricity (Earth eccentricity=1)	5.471
Mean surface temperature	-60 ° C
Mean surface pressure	6 mb
Mean surface gravity	372 cm/sec^2
Mean orbital velocity	24.1 km/sec
Mean density	3.933 gm/cc
Visual albedo or reflectance	0.159
Tilt of axis	25°
Sidereal day	24 hr 37 min 22.663 sec
Sidereal period	687 mean solar days
Mean distance from the Sun	227,941,000 km

The analogous right-handed amino acids (that may be similarly adsorbed by the surface of right-handed quartz), presumably may not have benefited from the small advantage due to the weak interactions in favor of the left-handed amino acids. Hence, it is forced upon us to discuss possible physical mechanisms that, while acting on both left-handed and right-handed amino acid crystals, may act as amplification mechanisms on chirally-pure samples.

Possible relevance of quartz in chemical evolution

The abundance of quartz and its presence in a variety of conditions typical of the Archean, militate in favor of the possibility that quartz may have provided early chirally-pure environments for amino acids (cf., Chapter 4). We shall consider a few arguments that favor a physical mechanism in which left-handed quartz is the first step in the pathway that leads to the observed homochirality of protein amino acids: the environment in which quartz is found is ubiquitous. Although quartz is one of the thousand silicate minerals that make up the bulk of the Earth's outer crust, it is however one of its commonest minerals (12 % by volume).

With the advantage of the experience gathered in the Martian space missions, we are in a position to discuss the main characteristics of the Martian soils.

TABLE 7.2: A selection of some of the missions sent to Mars.

Mission	*Launch date*	*Results*
Mariner 4	28/11/64	Fly by on 15/7/65 (First close-up images)
Viking 1	20/8/75	Landed on 20/7/76 (search for life)
Viking 2	9/9/75	Landed on 3/9/76 (search for life)
Pathfinder	4/12/96	Landed on 4/7/97 (First rover mission)
Global Surveyor (orbiter)	7/11/96	In orbit since 11/9/97 (mapping of the surface)

Quartz is known to crystallize directly from igneous magma (i.e., from molten rock material). It occurs in deep-seated coarse-grained ('plutonic') rocks such as granites, as well as in volcanic rocks such as rhyolites, which are igneous rocks containing crystals of quartz.

There is an additional reason to focus our attention on quartz. This mineral is stable in sedimentary conditions: it may occur in marine and desert sands. However, quartz can also occur as cement in compacted rocks such as sandstone in which quartz appears as rounded grains between the diameters of 0.06 and 2 mm, with a variable content of other mineral grains.

What seems even more pertinent is that quartz is also stable under extreme conditions of pressure and temperature. Precisely a high-pressure, high-temperature form (coesite) is found in meteorite impact rocks. Another form (stishovite) is found to occur naturally in meteorite craters, such as Meteor Crater, Arizona. Hence the probable presence of quartz in favorable environments during the heavy-meteorite impact-period (the Hadean) may have been a factor influencing the origin of life in the selection of homochirality, which has been perseveled as a distinctive feature of the earth's biota.

Furthermore, all terrestrial planets have a crust similar to the silicate crust of the Earth. In Tables 7.3 and 7.4 we show the presence of the elemental composition of Mars soil as well as the minerals that are on its surface.

In the case of Mars, for example, samples ('fines') have been analyzed by the Viking missions (cf., Figure 7.1). It was demonstrated that quartz is a prominent mineral in the fines that were retrieved from both the Chryse and the Utopia Planitiae [3].

The Viking missions (cf., Table 7.2) were able to retrieve sufficient information to make some very preliminary considerations in this respect. In the tables 7.3 and 7.4 we gathered some of the relevant information for the question of chemical evolution that was obtained from the Viking mission.

These elements make up compounds that are of interest in the theoretical and experimental discussions on the subject of chemical evolution and the origin and evolution of life in the universe. In particular it is of some interest to follow up some presumed implications (still to be proved by completely convincing arguments) of quartz in the Martian surface.

The Viking landing sites of the US mission of 1976 were two, the *Chryse Plains* located at 20 ° N, 40-50 °W and the *Utopia Plains* at 47.89° N, 225.86 ° W. (The usual convention is that the prime meridian is defined by a small crater named *Airey-O* , i.e., this location defines longitude).

Fig. 7.1: The 4000-km long Vallis Marineris is shown near the equator in an image taken by the Viking spacecraft (courtesy of NASA).

TABLE 7.3: Elemental composition of Mars soil [3].

Element	Formula
Silicon	Si
Iron	Fe
Magnesium	Mg
Calcium	Ca
Sulfur	S
Aluminum	Al
Chlorine	Cl
Titanium	Ti

TABLE 7.4: Minerals in Martian samples [3].

Mineral	Formula
Quartz	SiO_2
Feldspar	$KAlSi_3O_8/CaAl_2SiO_3$
Pyroxene	$MgSiO_3CaSiO_3$
Hematite	Fe_2O_3
Sphene	$CaTiSiO_5$

The Gas Chromatograph Mass Spectrometer (GCMS) test for organic molecules was the specific experiment responsible for the consensus that the Viking Mission found no persuasive evidence for life at Martian surface in the two locations chosen for the landing. GCMS results indicated no organic matter on the Martian surface within the sensitivity of the instrument winch was a few parts per billion. In fact, in the whole extensive area denominated the *Oxa Palus* region, extending from the equator to a latitude of 25° north, the Chryse Plains were chosen out of all the possible locations in this area only for safety considerations. (The Earth analogy would be to safely land the spacecraft in the Sahara desert to look for life, rather than in the more favorable environment for living organisms located at the Amazon River basin.)

The question of the origin, evolution and distribution of Martian life

The question of the possible existence of microorganisms in the solar system has been raised in the past, even independently of the Viking results, which we have already referred to in Table 1.1 [4]. Thus, we discuss here a possible solar-system environment in which a program for the search for eukaryotes may be relevant [5].

Our nearest-neighbor planet is a candidate for having supported life in the past. We cannot exclude its presence in some isolated environments. The possibility of extending the biosphere deep into the silicate crust in another terrestrial planet (Mars) deserves special attention [6]; this will be considered below. The present status of the search for life on Mars consists essentially of three Viking results:
• The nature of organic molecules on the surface.
• The presence of objects that may suggest living organisms or fossils through their motion or appearance, respectively.
• The presence in the soil of factors that may suggest metabolic activity.

The first aspect of the research was addressed by means of a very careful chemical analysis at the two landing sites. The second question was investigated by means of the camera borne by the lander spacecraft. Finally, the third question was addressed by means of three experiments (cf., Table 7.5).

However, UV radiation in the absence of a large fraction of oxygen in the Martian atmosphere (i.e., an ozone layer) prevents the possibility of life on the surface of that planet since, at least on the planetary surface, there are no evident UV defense mechanisms.

Not only have the biology experiments been repeatedly questioned, particularly in Japan, but also the possible existence of carbon compounds, including amino acids as well [7].

The Japanese group addressed the question of the possible existence of organic molecules in the polar caps and in permafrost elsewhere as a result of the interaction of cosmic rays and CO_2 in the Martian atmosphere. One possibility for living microorganisms surviving in Mars is represented by life underground, as a layer of permafrost could serve as the necessary UV defense mechanism [8].

This question seems pertinent to exobiology, since we cannot exclude at present that the organisms that have been found to inhabit deep in the silicate crust of the Earth may have been deposited with the original sediment.

Life may have evolved during an early 'clement period' that may have occurred in the Noachian Epoch, or Early Hesperian in Mars stratigraphy, according to the standard terminology [9]. Possible candidates for sites in which life may have evolved are located in the Tharsis region in the northern hemisphere, where volcanic activity has taken place since, by analogy with the Earth, the heat from underground magma may have produced hot springs, which are known to be possible sources of hyperthermophilic microorganisms. Knowledge of these possible locations raises the question whether life may have survived till the present confined to regions where pockets of liquid water may occur.

TABLE 7.5: The question of life on Mars.

Experiment	Results
Test for signs of photosynthesis or chemosynthesis induced by samples from the soil (the 'Pyrolytic Release experiment').	Small incorporation of CO/CO_2 into organic compounds.
Measurement of any gaseous products from a soil sample (the 'Gas-Exchange experiment').	Initial rapid release of O_2; slow release of CO_2 and N_2.
Search for the release of radioactive gas when the soil sample was exposed to a radioactive organic nutrient solution (the 'Labeled Release experiment').	Initial rapid release of labeled gas, followed up by slow release.

Has there been single-celled eukaryotes on Mars?

These considerations lead us to the conclusion that at present we cannot exclude the possible presence of single-celled eukaryotes in the Martian fossil record. Surface conditions on Mars are generally regarded not favorable for life.

The reason is mainly due to the extreme conditions compared to those enjoyed by the Earth biota. This withdraws the possible biotopes to deep underground; some estimates allow the possible Mars biota several kilometers beneath the surface, where there may be some liquid water.

There are several reasons why Mars may have experienced a more rapid environmental evolution towards an atmosphere rich in oxygen that may have favored for a restricted period a rapid pace of evolution of the background population of prokaryotic cells towards eukaryogenesis. As we have seen in Chapter 4 such an environment is favorable to the first appearance of the eukaryotic blueprint. On Earth eukaryogenesis occurred as far back as 2 Gyr BP, according to the micropalentological data.

Clearly such an "Eden" may not have lasted for long on a geologic time scale. At most we should expect some interesting work of "exopalentology" [10]. In Table 7.6 we have enumerated some favorable factors [11]. We have to wait for missions to Mars planned for the near future, particularly for the sample-return mission.

Was there liquid water on Mars?

The water inventory on Mars has been estimated from morphological evidence. An evident source of water ice is the north polar cap, which also contains dust ('dirty water') and carbon dioxide. The altimeter of the Mars Global Surveyor (MGS, cf., Fig.7.2) has made a map of the northern polar ice cap (cf., Fig. 7.3).

Figure 7.2:   A composite image of the orbital probe Mars Global Surveyor over the extinct volcano Olympus Mons, the largest volcano in the Solar System, 27 km high (courtesy of NASA).

TABLE 7.6: Factors favoring rapid evolution of the Martian environment [11].

Martian process compared to Earth	Comments
Some signs of ancient magnetic stripes have been discovered by the Global Surveyor. On Earth analogous magnetic banding is interpreted as evidence for plate tectonics.	One possibility is that the magnetic anomaly patterns detected may be some evidence for ancient plate tectonics.
Faster rate of hydrogen escape due to lower gravity.	Gravity on Mars = 0.38 gravity on Earth
Smaller oceans	Mars being a smaller planet would be less likely to intercept the trajectory of comets, hence less volatiles.

Figure 7.3: North polar cap (courtesy of NASA).

On the basis of these measurements it has been estimated that the diameter of this polar cap is 1,200 km and its maximum depth is 3 km. It is sufficient to cover the Martian surface to a depth of 10-30 m (or, equivalently about 4% of the total amount of water locked up in the ice of the Antarctic). An uncertainty is the unknown dust-to-water ratio. However, it is possible that liquid water existed in ancient times on the Martian surface, rather than simply water ice.

Indeed, the 'outflow channels' (surface structures measuring 10 km or more in width and hundred of kilometers in length), are known form aerial photographs; their slopes can be measured in terms of a few kilometers along their length. The origin of these features has been traced back to the catastrophic release of liquid water from the interior of Mars. We can estimate when liquid water flowed from a fairly accurate 'chronometer': the counting of craters on its surface. In the most remarkable outflow channels, which are in the northern hemisphere and drain into the Chryse basin, the crater count points towards an age of 3.5 Gyr BP.

The total amount of water estimated to have flowed along these channels is equivalent to covering the entire surface of Mars to a depth of 35 m. Less dramatic morphologic features that also were produced by water flowing on the Martian surface are the so-called *run-off channels* [12].

These structures are valley networks with dendritic drainage systems. They were produced by gradual erosion and slowly moving water. They are found in the heavily cratered terrain (hence very ancient) of the southern hemisphere. Their age is compatible with the period that followed the end of the heavy bombardment period some 3.8 Gyr BP.   What fed these valley networks may have been rain water, or even glacial melt. This water activity is an indicator of warmer climate and thicker atmosphere, conditions that have led to calling this period the Martian 'Eden', mentioned above.

During this period, life may have originated on Mars at a time when the Sun was not as luminous as it is now. There must have been a green house effect due to more abundant atmospheric carbon dioxide.

A special comment must be reserved for the images of the MGS studied during the year 2000 which led Michael Malin and Kenneth Edgett, scientists responsible for the MGS camera, to infer that many of Mars's meteoritic craters were the site of lakes during part of their history. According to their analysis the craters that have been considered contain accumulations of sedimentary rock that are several kilometers thick. The rock is divided into strata similar in color and thickness throughout Mars. The general distribution suggests that the sedimentation process was a global phenomenon rather than the result of local events (cf., in addition Fig. 7.4).

This is some of the evidence to the present time for what we have called above the Martian 'Eden", when the Red Planet was covered to a large extent by liquid water. This, in turn, adds some support to the possibility that life originated on Mars, or at least, that as a consequence of some large impact on the surface of the Earth, terrestrial life traveled to Mars during the Archean, closer to the period of heavy bombardment (cf., Chapter 3 for an account of the SNC meteorites that may provide examples of this possibility).

Unfortunately, we must leave the reader at the very frontier of current knowledge. It will be left for him to follow up ongoing discussions that concern the presence of water on the past or present of the Red Planet. One of them is the idea that Mars may at one time have had an ocean [13]. This tentative conclusion follows form data that has been gathered with instruments on MGS by James Head and his collaborators. The idea behind their work is that if an ocean existed at one time in the past, there should remain traces of a shoreline with an elevation that remains relatively constant.   At the same time the plains below should be smooth, due to a normal sedimentary process that oceanographers already have observed on our own planet. Other scientists have argued against this possibility. Another idea that will have to be evaluated is whether the liquid that modified some of the Martian surface was indeed water, or some other fluid.

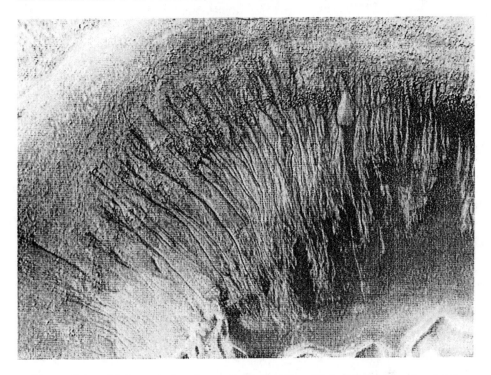

Figure 7.4: A high resolution image from the Mars Global Surveyor of a detail of a small 7-km crater inside the Newton Crater. It has been hypothesized that the channels that seem to flow down-slope could be the consequence the outflow of water from an underground layer (courtesy of NASA).

Mars Express

The ESA Mars Express mission [14, 15] is a European orbiter and lander mission, whose main objective is to study the interior, surface and atmosphere of Mars. The estimated launch is planned for mid 2003. from the Baikonur launch pad in Kazakhstan on board a Soyuz launcher. The mission will include various experiments. From the point of view of astrobiology the main highlight will be the Beagle 2 lander that will descend onto the Martian surface by parachute form the orbiter. This is a proposed lander whose main goal will be the search for organic material on the Martian surface and beneath it. Beagle 2 will also conduct miscellaneous studies of organic chemistry and mineralogy.

The landing mission will be provided with a rover that will be capable to search for samples underground and on rocks. The isotopic composition of the samples will also be studied. Beagle 2 is named after the ship in which Darwin and Robert FitzRoy sailed during the years 1831-1836 when he developed his ideas about evolution. No further mission to Mars in the near future has astrobiology as its main objective.

The future of humans on Mars

At present the Martian environment is unsuitable for any organisms to survive. However, the idea of sending humans to Mars is discussed with increasing intensity [16]. It is possible that with present technology, the main space agencies may send humans to Mars within the first two decades of the present century. The costs would not be prohibitive, if a new tendency is adopted to process the rocket propellants with Martian volatiles. It is expected that humans can transform the planet in the long term. After the experience of the Pathfinder, robotic and human exploration are feasible. On these bases, a program for planetary engineering may be possible too. However, beforehand we still have to make sure what are the risks of prolonged exposure to radiation as well as the effect of long-term weightlessness.

As to the usefulness of extending the Solar System exploration beyond robotic missions fortunately we have the experience of the Apollo Program in the late 1960s and early 1970s: That valuable effort showed the importance of having the missions with astronauts, instead of robots. The human presence is crucial for facing unforeseen problems such as repairing equipment and selective choice of rock samples.

Human assistance is also essential for planetary engineering. This activity has also been referred to in the literature as 'terraforming'. In this case humans will have to face two essential problems: to raise the level of suitable volatiles, and global warming. Current thinking makes us optimistic in this respect.

8

On the possibility of biological evolution on Europa

The satellites of the planet Jupiter were discovered early in the 17th century. The most intriguing possibilities for detecting extraterrestrial microbial life lie within the Jovian system, particularly in Europa, the second of the Galilean satellites.

The discovery of Europa

On January 10, 1610 while pointing his telescope at Jupiter Galileo Galilei, a lecturer of Mathematics at the University of Padua [1], observed three objects which he interpreted to be stars. He further noticed that two of these stars were to the east of the planet, while the third one was to the west. During the next evening he returned to his telescope and remarked that all three stars were at the west of Jupiter.

Finally, during the night of January 13 he saw a fourth object in the vicinity of Jupiter. These evenings were the crucial period in which there were the first intimations from observations that there was a center of motion other than the Earth. This put an end to one of the ideas about the immovable heavens maintained in ancient Greek philosophy that had lasted into medieval times.

The German astronomer Simon Marius subsequently (1614) gave these four satellites names from Greek mythology, corresponding, in part, to the maidens that Jupiter fell in love with. Marius called the second innermost satellite 'Europa', a choice which derives from Greek mythology, rather than Roman mythology. Zeus, the Greek equivalent of the Roman god Jupiter, was attracted by the charms of several mortal women. Amongst them was Europa, the beautiful daughter of Agenor, King of Phoenicia, from whose union was born Minos King of Crete. Io was another mortal woman with whom Jupiter fell in love. Io later married Telegonus, a mythical King of Egypt. Io was later confused with Isis, the Egyptian divinity.

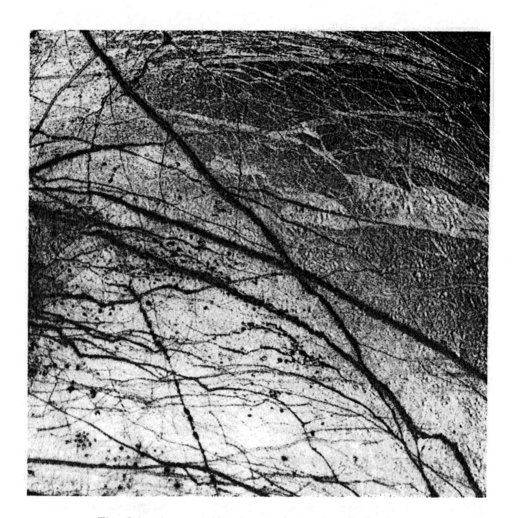

Fig 8.1: A composite image from the Galileo
spececraft of the Minos Linea region (cf., Table 8.2)
on the surface of Jupiter's moon Europa (courtesy of
NASA).

On the other hand Callisto was a minor divinity represented by a beautiful
maiden (a nymph) who, after being seduced by Zeus, was killed by the virgin huntress
Artemis, daughter of Zeus himself. Finally, the name of the fourth of the Galilean
satellites, Ganymede, unlike the other Galilean satellites, corresponds to that of a youth
taken to Olympus to become the gods' cup-bearer. The satellite Europa is a second
candidate for life in the solar system, after the more prominent planet Mars. This
follows from the indication that a large proportion of the spectroscopically detectable
material on its surface is water [2,3]. According to the results on the Jupiter system
obtained by the Voyager 2 mission, Europa is covered by a layer of ice (cf., Fig. 8.1),

under which there may be an ocean of water, whose temperature is 4 $^{\circ}$C. A comparison
with our own satellite is shown in Table 8.1.

TABLE 8.1: Europa, some information and statistics compared with the Moon .

Statistics	Moon	Europa
Mass (Earth = 1)	0.0123	0.0083
Diameter	3,476 km	3,138 km
Density	3.34 gm/cc	3.01 gm/cc
Radius of its orbit (both satellites have a small eccentricity)	Apogee: 406,740 km Perigee: 356,410 km	670,900 km 9.5 Rj (Rj = Jupiter's radius)
Bond albedo (or reflectance)	0,073	0.64
Surface composition	Solid	Water Ice
Surface gravity (Earth = 1)	0.165	0.135
Escape velocity	2.37 km/sec	2.02 km/sec
Orbital eccentricity	0.06	0.009
Sidereal period of revolution	27.32 days	3.55 days
Sidereal period of rotation	27.32 days	3.55 days

The great excitement that Europa has raised is due to the possibility of finding traces of life on that satellite. The features in Europa are probably the artifact of tectonic deformation of the crust. For comparison purposes, it is useful to list, in Table 8.2, some of the largest features on the surface of Europa.

By way of comparison, in Table 8.3, we point out two outstanding features of the Earth that are due to biological sources. This only serves to illustrate the fact that biogenic sources are known to influence the surface of the Earth and its oceans in a scale comparable to the largest geologic features known to exist in Europa.

Table 8.2: Some outstanding features of Europa [4, 19].

Feature	Details
Possible global ocean (but not proven)	Estimated depth 10 - 100 km
Surface composition	Water ice
Induced magnetic field and presence of hydrated salts	These physical and chemical properties support the presence of a global salty water ocean
Surface features (linea, an example of which is given in Fig. 8.1)	These features extend from about 150 to 2,700 km

From the similarity of the processes that gave rise to the solid bodies of the solar system, we may expect that hot springs may lie at the bottom of the ocean. It has been assumed in the past that Jupiter's protonebula must have contained many organic compounds (cf., Chapter 2, "Origin of the Jovian Planets"). Organisms similar to thermophilic archaebacteria could possibly exist at the bottom of Europa's ocean [5].

However, given the incomplete understanding of the evolution of early life on Earth, at present we should include eukaryotes as well amongst the possible components of the Europan biota. We may add that up to the present the divergence into the three domains, arising from the evolution of the earliest ancestor of all life on Earth, is not well understood. Indeed, plate tectonics has obliterated fossils of early organisms from the crust of the Earth, which is the only record available to us the evolution of early life on Earth.

The Galileo Mission

The Galileo Mission began to be planned in the 1970s, as the Voyager missions were being launched. The mission had been proposed to NASA by James Van Allen with the intention that some 12 orbits to the Jovian system be undertaken. Eventually, after some initial delays due to the Challenger accident, the mission took off and arrived at the Jovian system in December 1995 (cf., Fig. 8.2).

Galileo then released a 340 kg probe into the Jovian atmosphere. There were six instruments which were able to relay data to the orbiter. The lifetime of the probe was one hour. An important discovery was that the abundance of elements in Jupiter does not reflect the solar abundance, as might have been expected from the condensation of the solar nebula (cf., Chapter 2, "Origin of Jovian Planets").

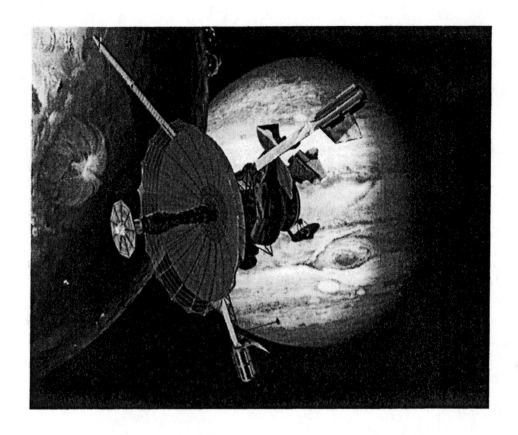

Figure 8.2: Galileo encounter with Jupiter's satellite Io (courtesy of NASA).

One possible conclusion is that since its formation, Jupiter's atmosphere has been considerably altered by the influx of small bodies, both comets and meteorites.

Table 8.3: Some large biogenic features on Earth.

Biogenic feature	Size (km)	Location
Great Coral Reef	2000	Australia
The Great Sponge Reef (fossilized)	3000	Europe

A year later (on 19 December 1996), the fourth orbit of the satellite Europa passed about 700 km from its surface and the data were processed the following month. The ice-cover of Europa was photographed in great detail. The question of the existence of an ocean is plausible, but further work is still necessary.

Tentative inventory of organic elements in the conjectured Europan ocean

Under the hypothesis that the Galilean satellites were of matter similar to the carbonaceous chondrites, it is possible to reach some useful estimates of their total 'volatile content' (substances with a relatively low boiling temperature). For this purpose consider Table 8.4, where the approximate correlation of the densities of the chondrites and satellites is given.

TABLE 8.4: Comparison between densities (g/cm^3) of the Jovian satellites and carbonaceous chondrites. (Adapted from ref. 5; cf., also ref. 11.)

Jovian satellite Name (density)	Carbonaceous chondrite Type (density)
Io (3.57)	III (3.5)
Europa (2.97)	II (2.5-2.9)
Ganymede (1.93)	I (2)

On the other hand, Table 8.5 suggests that CII carbonaceous chondrites have a sufficiently high fraction of water (13.35 %) to make up the volume required to fill up the presumed Europan ocean, provided that planetesimals of this composition made up the main part of the proto-satellite during the process of accretion (cf., Chapter 2, "Origin of the Satellites of the Jovian Planets").

Hence, in the formation of Europa from the Jovian protonebula, thermal heating of the initial CII carbonaceous chondrites would have led to the solution of biogenically important elements in the ocean (cf., Table 8.5: Mg, C, S). Table 8.5 also suggests that there is a substantial amount of carbon in Europa.

Habitability of Europa

In the present search for life in the solar system there are two prominent candidates: The first one is Mars, but the Jovian satellite Europa is a second environment in our own solar system in which we have proposed to test the hypothesis of the ubiquity of eukaryogenesis [6, 7]. This is briefly illustrated in Table 8.6:

TABLE 8.5: Chemical composition of carbonaceous chondrites, i.e., meteorites made primarily of silicates, but including some volatile elements (adapted from ref. 5).

	SiO_2 (%)	MgO (%)	C (%)	H_2O (%)	S (%)
Type I	22.56	15.21	3.54	20.08	6.20
Type II	27.57	19.18	2.46	13.35	3.25
Type III	33.58	23.75	0.46	0.99	2.27

TABLE 8.6: The question of life on Europa.

Voyager 2 Mission and observations from Earth.	Conclusions from the common origin of the Solar System.
Europa is slightly less dense than the Moon (3.04g/cm^3).	From typical Earth- crust rocks (3 g/cm^3), Europa may be inferred to be made mainly of rocky material.
Covered by a smooth layer of ice criss-crossed by a pattern of long cracks.	The absence of mountains and craters suggests it to be covered by water.
Under the surface there may be an ocean of water, whose temperature is 4°C.	Hot springs at the bottom of the ocean may imply the presence of microorganisms.

The existence of internal oceans in the Jovian satellites was conjectured as early as 1980, when Gerald Feinberg and Robert Shapiro speculated on this possibility in the specific case of the satellite Ganymede; more recently, Shapiro has further refined the Feinberg-Shapiro hypothesis by suggesting the presence of hydrothermal vents at the bottom of the ocean in Europa, in analogy with the same phenomenon on Earth [8]. In fact, for these reasons Europa is a candidate for a program for the search for extraterrestrial eukaryotes.

Search for biological evolution in the Solar System

Within the present program of exobiology in solar system exploration, the time is ripe to emphasize the life sciences, in addition to the more evident geophysical sciences. This should allow us to test the hypothesis of the ubiquity of eukaryogenesis in the case of the Jovian satellite Europa [9]. A second candidate is Mars. Previous work has suggested that there are at least two sources of heating of the satellite Europa independent of the Sun. They are tidal and radiogenic heating [10]. Of these two forms of heating, on Earth we are familiar with radiogenic heating, which is a consequence of the heat produced by the accumulation within the Earth crust of radioactive compounds.

The new factor in the Jupiter system is that unlike the Earth, Europa is influenced by its two neighboring satellites: the very volcanic satellite Io, which is closer to Jupiter and the giant satellite Ganymede, which is even larger than planet Mercury. The dynamics of this three-body system keeps the satellites from perfect circular orbits. The consequence is that the eccentricity of the Europan orbit varies significantly the vicinity to its giant planet thereby creating thermal gradients, which we have called 'tidal heating'.

Fortunately, we have some clear observational evidence of this form of heating, since Io, its nearest Moon-sized neighbor, is not covered with ice like Europa itself, neither has it a thick atmosphere such as that of Titan, the saturnian satellite.

It is possible then that a Europan internal ocean has consequently been formed (underneath a relatively thin ice cover) through dehydration of silicates [11], the heating source due to tidal heating is expected to be of the order of 52 $erg\ cm^{-2} s^{-1}$ with an additional 8 $erg.cm^{-2} s^{-1}$, due to radiogenic heating. For the above reasons it was estimated that the temperature underneath the icy crust could be $4^{o}C$.

From the similarity of the processes that gave rise to the solid bodies of the Solar System, we may expect that hot springs may lie at the bottom of the ocean. The main thesis of the proponents of the existence of a Europan biota is that, as Jupiter's primordial nebula must have contained many organic compounds, then possibly, organisms similar to thermophilic archaebacteria can evolve at the bottom of Europa's ocean.

The previous argument correctly pointed out that the most important requirements for the maintenance of life in Europa are the above mentioned conditions (liquid water, an energy source and organic compounds). However, the analogous Earth ecosystems considered have usually excluded eukaryotes.

The case of the acidophilic and thermophilic algae [12] is a warning that we should keep an open mind while discussing a possible wide range in the degree of evolution of Europan biota.

Additional factors arising from current experience with eukaryotes may contribute to clarify the case for not excluding nucleated cells from the microorganisms to be searched in new solar system environments; three arguments militate in favor of the ubiquity of eukaryogenesis in other ecosystems of the Solar System.

Is eukaryogenesis ubiquitous?

Some work [13, 14] suggests that protists co-existing in symbiosis with some prokaryotes may have proteins typical of eukaryotes. In addition, certain experiments with microorganisms [15] further suggest that the identification of eukaryotes at the morphological, or molecular levels during the process of fossilization may lead to confusing them with prokaryotes. Such difficulties suggest that exploration of Europa should not be confined to the possibility of designing equipment capable of recognizing exclusively analogues of the lowest domains of Earth biota (Archaea and Bacteria).

Consequently, we have proposed that tests for recognizing eukaryotes should also be envisaged in future missions [16]. We do not think it is appropriate to restrict the objective of future landing missions on the Europan ice exclusively to the search of prokaryotes, or their metabolites. If *living microorganisms* are found either in Europa (or in Mars), one may adopt a gene-centered approach [17], in order to search for cellular replication in relation to a delay in replication of chromosome segments. Such a phenomenon would confirm that the chromosomes maintain the condensed state during interphase, a eukaryotic hallmark. If frozen or fossilized organisms are encountered, then we need to consider [18] the differences between primitive eukaryotes and members of the other two domains.

What are the requirements for the maintenance of life on Europa?

In Chapter 2 we pointed out the similarity of the processes that gave rise to the planets and satellites of the inner and outer solar system. There is some evidence that satellites of the giant planets of the outer solar system are exposed to tidal heating, which as mentioned above due to the vicinity of a satellite to a giant planet, heating may be produced because of repeated stressing arising from orbital motion in the planetary field.

This is most clearly illustrated with the old Voyager images of the Galilean satellite Io. In the later visits to other outer solar system satellites during the 1980s, Voyager also discovered evidence of tidal heating in Enceladus, a Saturnian satellite, and also in Ariel, a satellite of Uranus. In these cases the images show that there has been resurfacing of these satellites since the period of heavy bombardment some 4 Gyr BP.

For these reasons it is reasonable to expect the presence of hot springs at the interface of the Europan silicate-core surface and the ice mantle. In fact, tidal heating in Europa is reasonable and its mantle of water ice can be partially liquid:

Europa experiences tidal stresses as it moves in its orbit, which is located between the orbits of Io and Ganymede. The multitude of linear markings on the Europan surface bear witness to such stresses.(cf., Table 8.2).

As we said above, the divergence observed in the terrestrial biota into the three domains is not well understood. Several factors arising from current experience with eukaryotes may contribute to clarify the case for not excluding these microorganisms from the microorganisms to be searched for in new solar system environments, including Europa.

Although it may seem unlikely for eukaryotes to have evolved in the environmental conditions currently hypothesized for Europa, at present we lack experience with parallel Darwinian evolution in extraterrestrial environments. This induces us to stress that present and future efforts should not be confined to the possibility of designing equipment capable of recognizing exclusively archaebacteria in missions to Europa which are envisaged for the period subsequent to the next orbital mission to the Jovian system: the 'Europa Orbiter' (cf., Fig. 8.3).

At the time of writing the launch date is likely sometime in the next ten years. The main objectives of the Europa Orbiter are to determine definitively whether or not Europa currently possesses an ocean of liquid water beneath its surface, to characterize Europa's subsurface structure, and to pave the way for subsequent exploration.

If an ocean is detected, other missions are likely, depending on the thickness of the ice layer. A lander is a possibility, which could look for evidence of life while sampling the surface composition and possibly performing seismic studies of the subsurface structure. Subsequent missions, which were originally called the hydrobot-cryobot mission may try to melt down to the ocean or even perform a sample return [19, 20].

Searching for microorganisms on Europa and terrestrial ice

As we have seen above, one of the most likely places in the solar system for the existence of extraterrestrial life forms is the Jovian moon Europa. We have discussed above the possibility that a volcanic-heated ocean exists underneath Europa's icy surface. After a first reconnaissance is made of Europa and it is determined that an ocean does exist under the ice, *in-situ* measurements will be needed to directly explore the Europan ocean and the ice that lies above it.

In order to make quantitative measurements of the Europan environment, a lander spacecraft was proposed a few years ago, which would be capable of penetrating the surface ice layer by melting through it [21]. This vehicle, dubbed a "cryobot", was suggested to carry a small deployable submersible (a "hydrobot") equipped with a complement of instruments (cf., Fig. 8.4). The cryobot is an adaptation of previous melter probes designed for terrestrial ice exploration; the hydrobot would draw on previous submersible experience from the oceanic research.

The proposed work describes a preliminary design of this mission. The question of instrumentation requirements for such a pair of exploratory vehicles is a fascinating problem, which still requires much research.

The design of an instrument package to search for life across the wide range of thermal and pressure environments expected on Europa, raises some issues in sample handling, and long-term reliability. Finally, this mission may present opportunities for performing high-value science on Earth, particularly in Antarctica, in order to test these instruments.

Testing for evolutionary trends of Europan biota

Given the possibility of designing an advanced lander mission that may melt through the ice layer above the putative Europan ocean, so as to deploy a tethered submersible, the question arises as to which suitable complement of instruments may be developed for testing for the presence of biogenic activity (to be discussed elsewhere). A second question concerns the parameters to be searched for.

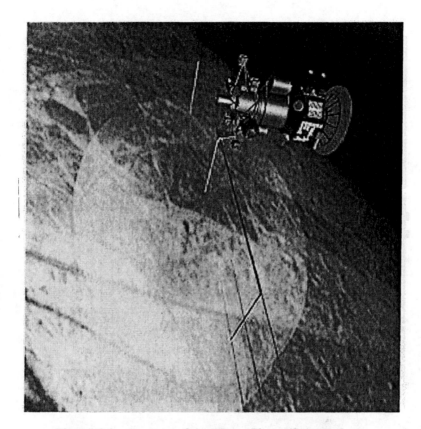

Figure 8.3: As part of its Outer-Planets/Solar-Probe project, NASA has begun development of the Europa Orbiter mission. The spacecraft, tentatively scheduled for launch in 2003, is under development at the time of writing (courtesy of NASA).

As the most likely biota that may be encountered in the Europan ocean are archaea [22], we argue that evolution should have occurred in Europa over geologic time, provided the ocean of liquid water has persisted for an equivalent period. We base this conjecture on the universality of natural selection and on the origin of archeabacteria at hydrothermal vents. The existence of life in Europa in the first place leads to a further question: whether sunlight is essential for the origin and evolution of life.

Can there be life without light?

The answer to this question is important for the possible existence of life on Europa, or any of the other iced satellites of the outer solar system, and elsewhere in the galaxy. Earth-bound eukaryotes depend on an the presence of oxygen in the atmosphere, which was in turn produced by prokaryotic photosynthesis over billions of years. A possible scenario favoring the existence of Europan biota decouples hydrothermal-vent systems from surface photosynthesis.

Figure 8.4: A hydrobot painting suggesting the search for microorganisms at the bottom of Europa's ocean. The submersible (hydrobot) is seen in the foreground (courtesy of Jet Propulsion Laboratory).

Indeed, experiments have already shown that chemical evolution leading to biological evolution is possible in conditions similar to those of hydrothermal vents [23]. Further, the delivery of amino acids at hydrothermal vents is possible, either by cometary or by meteoritic delivery [24]. Rather than prebiotic evolution, the genesis of *a primitive cell* in the deep ocean independent of photosynthesis is still a wider issue to be settled experimentally.

The possibilities of primary deep-sea or, alternatively, deep-underground evolution, are at present open questions. We may recall some related evidence against hydrogen-based microbial ecosystems in basalt aquifers, namely ecosystems in rock formations containing water in recoverable quantities [25]. This experiment raises doubts on the specific mechanism proposed for life existing deep underground [26]. However, the evidence refers more to the specific means of supporting microbial metabolism in the subsurface, rather than being an argument against the possible precedence of chemosynthesis before photosynthesis, which is really the wider and deeper issue to be settled. What remains to be shown in microbiology is that some barophilic and thermophilic microorganism has a metabolism that can proceed in completely anoxic conditions, deprived from carbon and organic-nitrogen derived from surface photosynthesis.

Thus, the case for life's origins, either through chemosynthesis first, or through a secondary reliance on photosynthesis at hydrothermal vents (by using oxygen dissolved in the sea-water), or deep underground, are still open questions. While this situation remains unsettled, plans for experiments have to be made by the space agencies, as the technological capability is consolidated for delivering equipment underneath the crust of satellites with iced surfaces, such as Europa.

Is there liquid water in Europa?

In 1976 the Voyager missions provided low resolution images of the surface of Europa. They showed a number of intersecting ridges and linae (cracks on the surface [27]). Besides, we learnt that craters were not abundant, suggesting that Europa has been geologically active until a relatively recent date (or, alternatively, there may have been 'resurfacing' from liquid water from below). To sum up, Voyager supported the intuition that planetary scientists already had. Such confidence was based on two facts. Firstly, from Earth-bound spectroscopy we knew that Europa is covered with water ice. Secondly, its density is not radically different from the Moon (cf., Tables 6.2 and 8.1); from a combination of these remarks it follows that Europa has a silicate core.

The Galileo mission has added much to the early insights we already had. One example is some form of 'ice-tectonics'. The Jet Propulsion Laboratory, which is handling the mission for NASA, have released some images which suggest that part of the surface is understood in terms of shifting plates of ice.

From all the information gathered from Voyager and Galileo reasonable guesses have been put forward in the sense that there may be a substantial amount of liquid water between the silicate crust and the iced surface. The trigger for the melting of the ice that we 'see' spectroscopically form the Earth could be tidal heating [28]. It may be useful to summarize the reasons for the excitement surrounding Europa's intriguing surface and mysterious interior. Liquid water is rare in the solar system. Water is universally required for the biochemistry of life being responsible, for instance, for the structure assumed by lipids in the lipid bilayer, which characterizes both eukaryotes and bacteria [29]. Some of the evidence for liquid water may be inferred form the linear features on the surface of this satellite (cf., Table 8.2). One particular case is provided by triple bands. Each structure consists of two dark bands separated by a bright intermediate ridge. The dimensions involved are colossal: the width of these triple bands is some 18 km; their length can extend for a thousand kilometers. These fractures on the iced surface may be the result of tidal deformation, which subsequently are filled in with water and silicates from its interior. After water and minerals are separated on the surface, freezing would result in water expansion to form the central ridge.

We may conclude that the possible presence of liquid water in Europa forces upon us the question of the possible presence of Europan life, as we have done in the present chapter. Galileo images and measurements have reinforced, but not yet proved, the existence of a Europan ocean [30]. One of the more significant measurements of Galileo has been followed up by Margaret Kivelson and collaborators. The method is based on the remark that in the presence of a magnetic field that varies with time, electrical currents are induced in a conductor (in the present case that interests us the example is water). Jupiter's axis of rotation is not lined up with its magnetic axis, the Galilean moons feel a fluctuating magnetic force. In the case of Europa an electric current is induced in the ocean. As a consequence, a magnetic field is induced, which is oriented in the opposite direction to that of created by Jupiter. Galileo has been in a position to measure the different magnetic fields, giving additional support to the presence of a sub-surface ocean.

What are the constraints on the putative Europan biotope?

A biotope is understood as a region uniform in environmental conditions and in its population of organisms for which it is the habitat. Speculating on the putative Europan biotope is not an idle exercise, for technology has reached a point in which funding seems to be the only barrier preventing us from a concentrated campaign of exploration of the putative ocean.

We have argued on the basis of the Galileo images that many hints already exist suggesting that part of the ice covering the silicate core is liquid water, forming an ocean, or lakes, analogous to those found in Antarctica (cf., Chapter 3). The same images allow us to go further: resurfacing of a large proportion of the Europan surface is a fact, since its craters are few in number. From the remarkable ones that do exist, Pwyll, Cilix and Tyre (cf., Table 8.2) provide further hints of its liquid-water interior. But resurfacing also hints at submerged geologic activity.

Europan volcanism may resemble terrestrial volcanism (this follows form the analogous formation of the terrestrial planets and the satellites of the Jovian planets, cf., Chapter 2). Earth-like geologic activity includes a candidate for the origin of life as we know it, namely, hydrothermal vents. It is clear that the chemical disequilibrium arising form thermal gradients present in the water circulating through the hot volcanic rock, could drive chemical reactions that lead to life [31].

The remaining question is whether an ample supply of organics could be found in the sea floor of the ocean. In Chapter 2 we sketched some aspects of the formation of satellites such as Europa. In fact, the carbon abundance of Europa should be correlated with solar abundance (cf., Table 1.2). This is further supported by evidence supplied by the carbonaceous chondrites (cf., Table 8.5). We have argued above that the Europan sea floor could be a suitable site for the evolution of life. So if archaea are possible dwellers of hydrothermal vents, there is no evident reason preventing the prokaryotic blueprint form evolving into the eukaryotic blueprint.

9

On the possibility of chemical evolution on Titan

The current interest in Saturn's satellite Titan, which is the second largest satellite of the Solar System, is mainly due to the last great mission of the last decade of the 20th century (Cassini-Huygens), an ESA-NASA collaboration that aims to drop a probe through the thick atmosphere of Titan.

The discovery of Titan

In 1655 the Dutch astronomer Christiaan Huygens discovered a moon rotating around the planet Saturn. The Italian-born astronomer of the Paris Observatory, Jean Dominique Cassini observed the gap in Saturn's rings (Cassini's Division) in 1675. The current joint mission ESA/NASA takes his name and that of Huygens. The 'Cassini' probe will explore the Saturn system and the probe Huygens will be dropped into Titan atmosphere (cf., below: The Cassini-Huygens mission).

Following the tradition that had started with the Jovian system, names for the system of Saturn were chosen from Greek mythology. In the case of the new Huygens satellite it was natural to return to the Titans, who formed the first divine race. (The name probably comes form the Cretan word for 'king'.)

In the beginning of last century (1908), the Spanish astronomer Josep Comas Solá from the Fabra Observatory in Barcelona, discovered that Titan was the first satellite in the Solar System to have an atmosphere.

The distinguished English physicist Sir James Jeans studied the problem of atmospheric physics of the bodies in the solar system. In the case of Titan he was able to determine the factors that might prevent atmospheric gases from escaping even a body as small as a satellite. To counteract the weak gravitational hold on the gases, Jeans established a limit for retaining the atmosphere, provided that the temperature was sufficiently low. The gases that are allowed in the Titan atmosphere include methane

117

(CH$_4$, since at the low temperatures in the Jeans calculation still remains in its gaseous state.)

For the discovery of methane in Titan's atmosphere, we had to wait till 1944, when the Dutch astronomer Gerard Kuiper added a second clue to the singular features of Titan. He discovered that its atmosphere contained the gas methane. It was soon realized that the reason for the sizeable satellite to hang on to an atmosphere was more due to its extremely low temperatures rather than by the force of gravity (confirming observationally the theoretical calculations of Jeans). Indeed, Kuiper identified the spectral signatures of methane at wavelengths longer than 6000 Å (there are two such lines).

Before proceeding to a more careful discussion of Titan the reader is advised to consider some of the satellite statistics in Table 9.1:

TABLE 9.1: Characteristics of Titan [2, 3].

Parameters	Values
Radius	2,575 km
Mass (Earth mass = 1)	0.022
Surface gravity (Earth gravity = 1)	0.14
Mean density	1.88 gm/cc
Distance from Saturn	1.226×10^6 km (20 R$_S$)
Orbit period	15.95 days
Obliquity	26.7 °
Surface temperature	94 K
Surface pressure	1496 mb

The Voyagers: The first missions to Saturn and Titan

The Voyagers consisted of two space probes sent by NASA in 1977 to explore the solar system, but the results for Saturn in particular are worthy of some comment. Voyager 1 reached Titan on 12 November 1980. Voyager 2 was also able to explore Saturn, but the distance of closest approach of Voyager 1 was a hundred times nearer.

Two features are remarkable in the Voyager 1 images [1] (cf., Fig. 9.1):

• There is a difference in brightness between the two hemispheres. It has been referred to as the north-south asymmetry. It is probably due to circulation of its thick atmosphere. Subsequent observations form the Earth's orbit, by means of the Hubble Space Telescope have shown that the asymmetry can change with time. Hubble was able to focus on Titan in 1990; but the observation of Hubble and Voyager were at different wavelengths.

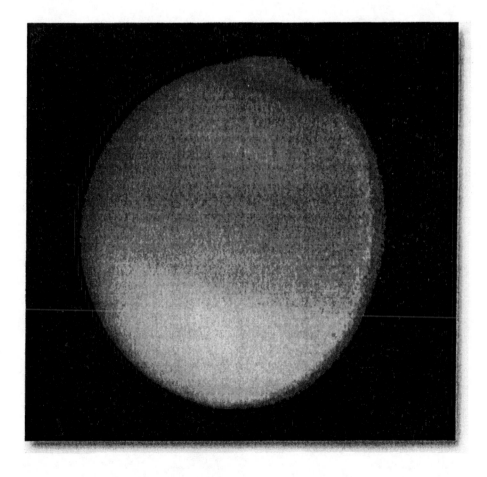

Figure 9.1: This Voyager 2 photograph of Titan, was taken on August 23, 1981 from a range of 2.3 million kilometers. It shows some detail in the cloud systems discussed in the text. The southern hemisphere appears lighter in contrast, a well-defined band is seen near the equator, and a dark collar evident at the north pole (Voyager image, courtesy of NASA).

• The other significant feature in the Voyager 1 images is that a 'cap' can be seen close to the north pole of Titan (strictly, the feature resembles a dark ring and is referred to as a polar hood). It covers a large part of the northern hemisphere from 70°-90° north latitude. Planetary scientists assume that this feature may be caused by either the lack of illumination in the winter or alternatively it may be caused by the dynamics of the atmosphere.

Titan and Triton

To appreciate Titan's features it is important to compare it with another satellite, of the outer solar system, the Neptune's moon Triton, particularly because they both have atmospheres, mainly of molecular hydrogen (cf., Fig. 9.2).

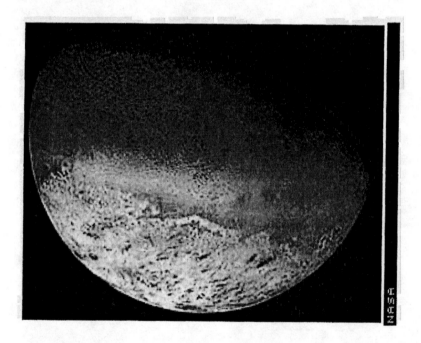

Figure 9.2: Triton the largest moon of the planet Neptune (courtesy of NASA).

In Table 9.2 we summarize the main physical parameters of both satellites. Triton is probably a captured object.

The formation of Triton would have been in the solar nebula, unlike the case of Titan, which is believed to have been formed in the disk of material that surrounded Saturn. Its density is the same as that of the two other giant moons of the outer Solar System, the galilean satellites Ganymede and Callisto.

TABLE 9.2: Comparison of the main physical parameters of Triton and Titan [1,4].

Parameters	Triton	Titan
Radius (km)	1,352	2,575
Mass / Mass of the Earth	0.004	0.023
Average density (kg/m^3)	2100	1880
Mean distance from the planet (radius of the planet R_P)	14.33 R_N	20R_S
Distance from the Sun (AU)	30	9.6
Period in orbit (days)	5.88	15.95
Composition of its surface	Ices of N_2, CH_4, CO, H_2O, CO_2.	Ice? Hydro-carbons?

Triton's rotation around Neptune is in a direction opposite to the planet's rotation ('retrograde motion'), an unusual case for our solar system, adding further support to the hypothesis of the captured-object.

The atmosphere of Titan

The atmosphere of Titan is one of its most interesting features. As we mentioned above, Kuiper's discovery of spectral signatures in the atmosphere correspond to methane. More recently and by means of the Voyager 1 spacecraft, the atmosphere of Titan has been demonstrated to be made mainly of nitrogen and methane with small amounts of hydrogen cyanide, cyanoacetylene, and other organic compounds. As we have seen in the previous section, from the Earth, Titan has the appearance of an orange haze, whose major components have been studied with various means, including the Voyager 1 flyby. In Table 9.3 we have listed some components of the atmosphere.

Carl Sagan and Bishun Khare in the early 1970s did a set of chemical evolution experiments in which they irradiated several methane-rich atmospheres with ultraviolet light or electrons. This led them to solids that were colored in a manner reminiscent to the color of Titan's atmosphere (reddish/brownish). This and many other insights coming from several chemical evolution laboratories elsewhere make the forthcoming Cassini/Huygens mission one of the astrobiological highlights of the near future.

TABLE 9.3: Major components of Titan's atmosphere [2].

Gas	Mixing ratio
Nitrogen, N_2	0.88-0.98
Argon, Ar	0-0.06
Methane, CH_4	0.017-0.06
Hydrogen, H_2	0.0006-0.0014

After a seven-year voyage, a probe, the Huygens, will be separated from Cassini (cf., Fig. 9.3); it will then be parachuted through the atmosphere (cf., Fig. 9.4), initially at a speed of some 20,000 km/hr. When Huygens reaches the surface, as shown in Table 9.1, it will be working at temperatures of -200 °C (94 K). The mission is equipped with a radar altimeter and imaging devices, in order to probe the surface of Titan as it approaches the satellite. For a fuller account we refer the reader to "The Cassini-Huygens mission", below. One of its goals is to identify *in situ* the atmosphere constituents, of which we have tabulated the main components in Table 9.2. Some of the minor components of the atmosphere are given in Table 9.4.

The origin of Titan's atmosphere

If we return to Table 1.2, we can verify that if Titan's atmosphere had been captured from the Saturn protonebula, the abundance of the element nitrogen (N) would have been similar to that of neon (Ne). The abundance of both of these elements is almost identical in the Sun. By implication, both N and Ne should have been equally abundant in the solar nebula.

We can verify from Table 9.3 some information on the elements in Titan's atmosphere. This data has been gathered since the time of the Voyager missions. We remark that since neon is known to be depleted with respect to nitrogen, it follows that Titan's atmosphere was not formed from the Saturn protonebula itself.

The most likely hypothesis for the formation of the atmosphere is outgassing from the materials that made up this giant satellite. The primitive atmosphere received subsequently contributions from comets coming form outside the Saturn system.

The tantalizing surface of Titan

A number of observations have provided much information on the nature of Titan [5]. From this work educated guesses can be formulated as to the nature of its surface, which is invisible from either the giant Hubble Space Telescope, or form the previous close encounter of the Voyager mission at the beginning of the 1980s. However, the Voyager mission was able to detect the exact measurement of the satellite radius. Previously what was known was the size of the satellite and its thick atmosphere.

The relevant experiment was named a radio occultation [6]: as the spacecraft went behind Titan the radio signal transmitted was interrupted exactly as its solid surface interrupted the transmission. From careful observation from the tracking station, the radius was obtained. (The atmosphere is not dense enough to stop the signal.) Coupled with some calculations, the radio occultation measurements of Voyager demonstrate that there is a depth of some 200 km from the surface to the visible limb. Once the radius is known, only one more parameter is needed to infer the density of Titan, namely its mass. This was also done by the Voyager mission simply by detecting the deflection of the radio signal sent by the spacecraft due to the gravitational interaction.

The result shows that the density is much lower than the corresponding value for the Earth (cf., Table 9.1). This is the basis for our conjectures regarding the nature of the surface of Titan [5]. Indeed, there must be materials on the satellite which will account for the much lower density of Titan.

A structure similar to the iron core of the Earth cannot be present. But if we consult the corresponding value for the Moon on Table 3.1 and recall that its constituents are predominantly silicates, we still can infer that to reach such a low value of density, as observed on Titan, lighter materials must be present.

In fact, we may conclude that a large quantity of lighter materials such as liquid or solid hydrocarbons must be present on the Titan surface. But we may ask, what might be the distribution of the liquid and solid hydrocarbons on the surface? Although at visible wavelengths the atmosphere is opaque, infrared observations carried out with the help of the Hubble Space Telescope show variations in surface albedo that are consistent with the presence of continents and oceans [7]. (At such low temperatures the oceans, if they exist, are most likely of liquid hydrocarbons, not of liquid water.) We may formulate conjectures as to the nature of the solid surface. For we must assume that the number of craters on the surface of Titan can be interpolated from the crater density observed elsewhere in the Saturn system in which so many satellites lacking an atmosphere have been photographed. Typically there should be about 200 craters larger than 20 km diameter per million square kilometers.

The exact configuration of craters has to await the landing of the Huygens probe (cf., section below). Very exotic landscapes can be conjectured in which large craters filled by liquid hydrocarbons may be present. Some of these ideas may actually be observed in the year 2004 with the landing of Huygens.

The Cassini-Huygens mission

This mission is one of the most advanced space efforts undertaken so far. We should mention, in addition to what was said above (and in Fig.s 9.3 and 9.4), the many insights to be gained into the nature of Titan. Outstanding amongst these is a scientific investigation of its atmosphere and surface; but Cassini-Huygens will also investigate the rings of Saturn, the magnetosphere of the giant planet; it will also study the surface of the icy satellites of Saturn [3,5].

TABLE 9.4: Minor components of Titan's atmosphere [2].

Gas	Mixing ratio
Hydrocarbons (at the equator)	
Ethane, C_2H_6	1×10^{-5}
Acetylene, C_2H_2	1.8×10^{-6}
Propane, C_3H_8	7.0×10^{-7}
Ethylene, C_2H_4	5.0×10^{-9}
Diacetylene, C_4H_2	1.4×10^{-9}
Nitriles	
Hydrogen cyanide, HCN	At the equator 1.4×10^{-7}
Cyanoacetylene, HC_3N	At the north pole: 10^{-8}-10^{-7}
Cyanogen, C_2N_2	At the north pole: 10^{-8}-10^{-7}
Oxygen compounds	
Carbon dioxide, CO_2	At an altitude of 108 km: 3.3×10^{-9}
Carbon monoxide, CO	
	At the troposphere: 6×10^{-5}

In fact, in June 2004 Cassini will begin to explore Saturn's 18-satellite system. The mission will last four years, and at the time of writing the many objectives of the mission have been carefully scrutinized. Not only is Titan a major puzzle for extraterrestrial chemical evolution, but the satellite Iapetus, with sharply different albedos on both hemispheres will be one of the objectives for clarification. The Cassini mission will explore the Saturn 'planetary system' for a period of four years (2004-2008). A variety of different experiments will be performed (cf., Table 9.5).

This rich system contains rings encircling the planet, many icy satellites, of which Enceladus is probably one of the most mysterious with evidence of some geologic activity in spite of its small dimension; the magnetosphere is also of considerable interest with the myriad particles.

Huygens is wrapped in a heat shield that will protect its equipment during the descent in the Titan atmosphere. Its viability was demonstrated in 1995 when a trial was performed in northern Sweden [8]. From a balloon carrying a full-scale model of Huygens

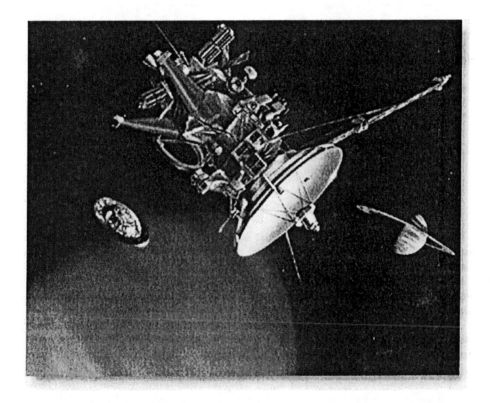

Figure 9.3: An artist's conception shows the moment the Huygens probe detaches itself from Cassini to begin its descent in Titan's atmosphere (courtesy of NASA).

in a gondola, the replica of the probe was dropped from a height of almost 40 km.

Once the process of shedding its shields and deploying the parachutes was completed, it was confirmed that all the instruments worked as they should in the mission in the year 2004.

TABLE 9.5: A sample of experiments planned for the Cassini-Huygens mission [3,5].

Cassini Instruments	Huygens Instruments
Optical remote	Surface science package
Fields, particles and wave instruments	Huygens atmospheric structure instrument

Microwave remote sensing
   Cassini radar
   Radio science subsystem
	Aerosol collector and pyrolyser

Can there be liquid water on Titan

We have seen in Chapter 8 that one of the key questions to study in the future is the possibility of searching for extraterrestrial microorganisms. From what we have discussed in this chapter, prokaryogenesis on Titan, the first stage of the evolution of life as we know it on Earth, seems a very unlikely possibility. Nevertheless, Titan still presents itself as the world most similar to the early Earth during its stage of prebiotic evolution before the dawn of life.

In Table 9.3 we have illustrated the very rich organic chemistry that we already know to exist on our sister world. For this reason the question of the presence of liquid water on Titan remains as a high priority in the questions to be answered in the future, even in the post-Cassini-Huygens era.

The question of the presence of liquid water on Titan is not irrelevant. From Table 9.1 the reader can infer that for such a low value of its mean density (0.56 of the Moon's mean density, cf., Table 3.2), a large proportion of its mass must consist of volatiles. Yet, independently, we do know that water is one of the abundant compounds in the universe.

Hence we expect a fraction of its mass to consist of water. On the other hand, we also know from Table 9.1 that the surface temperature of Titan is 94K. Most of its water content is expected to be sequestered in the form of water ice. Hence, a leading question for the future research to decide is if hidden underneath its thick clouds, Titan may have some volcanic activity that could lead to some permanent liquid water.

Nevertheless, independent of the presence of volcanic activity, liquid water may be present. The above comparison of Titan's mean density with that of a silicate body like the Moon suggests that perhaps 30% of its mass should consist of an ice mantle over the silicate core [9]. A few hundred kilometers beneath the surface, the likely prevalent physical conditions are: low enough pressure coupled with low temperature; yet, these conditions would be sufficient to allow a subsurface ocean of liquid water.

This would be a second example, much in analogy with the Europan ocean, which we have already discussed in Chapter 8. It should be noted that all the ice would probably not be all pure water, but a substantial fraction of ammonia, for instance, is also possible.

Episodes of liquid water can clearly happen due to the collision of meteorites and comets, as has been pointed out in the past [6]. It should be underlined that even though Io is relatively much closer to Jupiter ($6R_J$ in units of planetary radii) than Titan is to Saturn ($20R_S$), our inability to predict the effect of tidal heating of a giant planet on its satellite was recently demonstrated by the unexpected high temperature of silicate volcanism, as seen by the Galileo mission [10].

Figure 9.4: This artist's conception shows the Huygens probe as it descends in Titan's dense and murky atmosphere of nitrogen and carbon-based molecules (Photo ESA).

Indeed, the Galileo study showed that Io's erupting lava are probably hotter than the hottest temperature basaltic eruptions that occur on Earth. These uncertainties underline the fact that the question of the presence of liquid water on Titan is still an open question. Hence the question of whether there could be some form of life on Titan remains equally unanswered.

The possibility of life on Titan is remote

We should notice in passing that it is difficult to exclude *a priori* microbial life altogether, since the lack of atmospheric oxygen was overcome by early Earth methanogens, a form of extremophilic bacteria that can metabolize methane. Another factor that adds excitement to future research is the fact that only recently we have begun to realize that a large number of prokaryotes do live deep underground [11], a possibility that cannot be easily excluded in Titan if episodes of volcanic activity (and hence of liquid water) occurred earlier in its history, giving rise to isolated underground ecosystems. On Earth subsurface has been defined as habitats below 8 m and marine sediments below 10 cm. Circumstantial evidence suggests that the subsurface biomass on Earth is very large. Up to one million prokaryotes per *ml* have been found in ground water form deep underground.

However, the speculation of possible earlier life on Titan remains as a challenge that Cassini-Huygens will begin to unravel and possibly rule out. With our present state of knowledge life on Titan remains as a remote possibility to be excluded with the ever increasing amount of knowledge that will be available in the post Cassini-Huygens era.

Part II

BIOASTRONOMY:
THE STUDY OF ASTRONOMICAL PHENOMENA
RELATED TO LIFE

10

How different would life be elsewhere?

In the next two chapters we shall be discussing possible environments for life and intelligent behavior in the universe. In this context we shall also address the following question: How can we subject to experiment the existence of eukaryotes outside the bounds of their terrestrial evolutionary pathways?

Possible degrees of evolution of life

We wish to discuss in the first place the different positions that are possible regarding the question of extraterrestrial life. It is against this background that we should test the validity of the main conjecture of this book. We have understood sufficient aspects of the question of eukaryogenesis so that we may begin the search for life in solar system exploration with a more advanced approach. We will ask more than whether life is present in other environments such as the Mars permafrost or in the possibly molten ices of the Jovian satellite Europa. We wish also to know what is the *degree of evolution* of the putative extraterrestrial life that we are hoping to encounter in the various space missions.

This question is not just due to curiosity. We are approaching a unique moment in the development astrobiology with new vigorous efforts being followed by the main space agencies. Time and funding are crucial in this endeavor, so that if the question of the degrees of evolution can be faced from the start, we should make every effort to do so. We can only distinguish between alternatives by experiments either *in situ*, or by sample-return-missions.

We list a few questions that have to be considered:

• The more widely accepted belief on the nature of the origin of life is that life evolved according to the principles of deterministic chaos. Evolutionary developments of this type never run again through the same path of events [1]. The possibility that similar evolutionary pathways may have been followed on different planets of the Solar System has been raised [2].

Indeed, even if some authors may consider this to be a remote possibility, this work is based on the increasing acceptance that catastrophic impacts have played an important role in shaping the history of terrestrial life. Thus there may be some common evolutionary pathways between the microorganisms on Earth and those that may have developed on Mars during its 'clement period' (roughly equivalent to the early Archean in Earth stratigraphy). The means of transport may have been the displacement of substantial quantities of planetary surface due to large asteroid impacts on Mars.

• Regardless of the possibility raised above, many researchers still see no reason to assume that the development of extraterrestrial life followed the same evolutionary pathway to eukaryotic cells as it did on Earth. Moreover, a widespread point of view is that our ignorance concerning the origin of terrestrial life does not justify the assumption that any extraterrestrial life form has to be based on just the same genetic principles that are known to us.

In sharp contrast with this position, it has been argued that lessons should be drawn from principles of genetics [3]. Various approaches regarding the possible implications of our current understanding of life's origins have been suggested, but genetics provides a point of view that may be said to add a fourth way to the question of the nature of extraterrestrial life.

• We all agree that the final outcome of life evolving in a different environment would not be the same as the Earth biota, but new ground has been broken with the question used as a title for the present chapter, more precisely:

> How different would such alternative extraterrestrial evolutionary pathways be?

This has led to a clear conclusion that there is no reason for the details of the phylogenetic tree to be reproduced elsewhere. The tree of life constituted by all living organisms may be unique to our planet. On the other hand, there is plenty of room for the development of differently shaped evolutionary trees in an extraterrestrial environment, where life may have taken hold. But, in agreement with De Duve, it is possible that the selective advantage of certain directions, for instance in the pathway that led to brains on Earth (cf., Chapter 12) may have a high probability of occurring elsewhere in the universe [3].

Such an argument supports the initiative of going to the other extreme of evolution from microorganisms, where it is reasonable to search for the activities of intelligent behavior. In our extensive experience on this planet with several stages of evolution, living organisms that have brains, do have a tendency to communicate amongst themselves. In the next chapter we shall discuss the various efforts that have been followed up in this direction.

The search for other solar systems

As we have seen in Chapter 2, the Sun and planets condensed from a large rotating disk of dust and gas, themselves the product of condensation of the interstellar medium. The bottom line of this hypothesis is that most of the angular momentum of the parent solar nebula is presently concentrated in the orbital motion of the planets. For this reason, it is only reasonable to assume that the leading candidates for possessing planetary systems are single stars.

Besides, it is considerably difficult to infer their presence from the interaction of the planets with their parent star. For instance, the partial eclipse of a given star can occur if the ecliptic plane is in a favorable alignment with the plane of the planets that rotate around other solar systems.

It is instructive to consider a sample of stars, emphasizing some known to be somewhat similar to our own solar system, as they have Jupiter-like planets, although in several known cases these giant planets are in very small orbits around their stars; for instance, in the case of the first extrasolar planet discovered by Michel Mayor and Didier Queloz in 1995 [4].

In *Pegasi 51*, the Jupiter-like planet was only about seven million kilometers away. This raises the question as to the mechanisms that might either give birth to a giant planet close to the star or that would induce the orbit of a giant planet, such as our own Jupiter to decay into a Mercury-like orbit.

However, the fact that only Jupiter-like planets have been discovered since the first report of the *Pegasi 51* system, does not exclude the possibility that planets of smaller mass (with their corresponding satellites) will eventually be known to us when measuring techniques become more accurate. Such discoveries suggest that life has been provided with a variety of appropriate environments in our own galaxy; this is illustrated in Table 10.1. (In view of the well-established existence of many solar systems, throughout the present work we have only capitalized the phrase 'solar system' whenever we wish to emphasize our specific cosmic environment.)

In the 1970s Carl Sagan and co-workers used a computer program in order to assess what type of distribution one should expect in other solar systems [5]. However, the theoretical models did not anticipate the surprising results of the planets discovered since a planet was detected around *Pegasi 51*.

How are extrasolar planets found?

The key concept is that when a sufficiently large planet rotates around its star, say Jupiter rotating around the Sun, the gravitational interaction between planet and star is reciprocal and there is a mutual perturbation.

Given the development of astrometry to this date, the pair Jupiter-Sun produce a perturbation of the Sun that would be detectable within a radius of 100 light years. The star goes into a minuscule orbit, but not negligible. The stellar motion is referred to technically as the *reflex motion*.

There are two ways to determine the reflex motion [10]. We can first detect the change of velocity of the star with respect to the Earth. In physics the technique is known as the Doppler effect: as the star is moving towards or away from us, there will be a measurable difference in the wavelength of the light reaching us. This is a general phenomenon, which consists of the apparent change in frequency of a source of light (or sound, or indeed any wave motion) due to the motion of the source and observer.

The second means of identifying the reflex motion of the star is by means of astrometry (the subject which concerns itself with the observation of the position of celestial objects and its variation over time [11]).

Which are likely habitable zones?

The state of the art in detecting the reflex motion only allows one to detect the star motion due to the presence of a Jupiter-like planet. Indeed, the nine examples that we selected in Table 10.1 are all Jupiter-like planets. It is remarkable that the planets themselves that have been detected up to the present are the least likely to be habitable. However, these Jupiter-like planets could have Europa-like satellites (cf., Chapter 8), which could be likely candidates for harboring life.

This opens the debate as to what are habitable zones, a subject of great interest which we shall not develop in detail here. A habitable zone was understood as one in which the amount of stellar energy reaching a given planet, or satellite, would be conducive to the process of photosynthesis. (Clearly the amount of green-house gases, such as carbon dioxide, in the atmosphere are relevant too.)

Stars and their Jupiter-like planets

We have seen in Chapter 1 the importance of the HR diagram. Once again it is useful to the astrobiologist: the mass of the companion to the star that wobbles-its reflex motion- can be calculated [10] as follows: the mass of the given star (for which we want to ascertain whether it has a companion planet) can be determined from its temperature and from the total amount of light that it emits. The HR diagram correlates the three parameters: mass M, temperature T and total amount of light (E) that it emits. Hence, with only two measurements (E, T) we can infer M. Once M is known and the amplitude of the reflex motion is carefully measured (by means of the variation of the star velocity with time), the mass of the companion can be determined.

Jupiter in our Solar System is the largest planet, capable to introduce perturbations into the central star (the Sun) detectable form other worlds (cf., Figure 10.1).

So far all the extrasolar planets have Jupiter as a point of reference (cf., Table 10.1), some are the same size others are substantially larger, so much so that even the question of the definition itself of a planet is debatable (cf., next section). A complication worth noticing has already been made manifest in our Table 10.1, namely, if the velocity of the star is not precisely towards and away form us along the line of sight between us and the wobbling star, the best we can do is to determine the component of the velocity that is along the line of sight-this is the significance of the sin(i) factor in the above-mentioned Table 10.1. The bottom line of this comment is that the mass of the companions that are cited in Table 10.1 are really lower bounds or equal to the real mass of the companion.

What is a planet?

There are two cases in which the answer to the above question is not clear. In the outer Solar System the dividing line between a large Kuiper belt object and a small mass planet is rather blurred.

The case in question concerns the discovery during the 1990s of over 60 mini-planets, whose diameters range between 100 and 500 km. By simple extrapolation, it is

TABLE 10.1: Some examples of stars with confirmed planets in order of ascending mass up to a cut-off well below 10 Jupiter masses [6, 7, 8]. M_J is the mass of Jupiter; the results obtained are for the parameter M sin (i), where M is the new planet's mass, and i is the orbital inclination to our line of sight (this is an unknown parameter, and so the masses are maximum values).

Star	Distance from the Sun (pc)	Distance of its Jupiter-size planet: semi-major axis(AU); eccentricity	Size of its Jupiter-size planet (M_J)
51 Pegasi	15.4	0.051; 0.01	0.44
Upsilon Andromedae (*)	16.5	0.053; 0.03	0.63
Rho 1 55 Cancri (a double System)	13.4	0.12; 0.03 >4; -	0.85 >5
Rho Corona Borealis	16.7	0.23; 0.05	1.1
16 Cygnus B	~22	1.70; 0.68	1.74
47 Ursae Majoris	14.1	2.08; 0.09	2.42
Tau Bootis	~15	0.042; 0.00	3.64
70 Virginis	18.1	0.47; 0.40	6.84

(*) In the important case of Upsilon Andromedae, where a solar system has been identified by two groups of astronomers, we have only inserted information on the first of the three Jupiter-like planets that have been reported (the other two are of masses that double and quadruple the Jovian mass [9]).

now believed that in the region between 30 to 50 AU on the plane of the ecliptic (where the Kuiper belt extends), there might be up to 40,000 such objects. At the other end of the spectrum, when we consider the extrasolar planets that we have been discussing in this chapter, the question of the brown dwarfs has to be faced [12].

Indeed, the dividing line between a small brown dwarf and a large Jupiter-like planet is not easy to distinguish. In fact, a brown dwarf is an object formed in the same way as a star. However, the mass involved in the process of star formation is insufficient for triggering the thermonuclear reactions that are necessary for the birth of the star. The lower bound for star formation is about 80 Jupiter masses. However, the minimum mass is unknown for a brown dwarf.

Figure 10.1: Jupiter, the planet of our Solar System that serves as a reference for estimating the size of extrasolar planets mentioned in Table 10.1 (Voyager 1 image, courtesy of NASA).

Just like in many chapters in this book, the reader is invited not to be disappointed, but rather to wait anxiously for the technical developments of the next decade, when direct observation of the extrasolar planets might settle the question one way or another.

The orbital telescope called the "Next Generation Space Telescope" (NGST) is under construction. Its launch is scheduled for 2007. Although it will carry a high resolution infrared observatory, it is still not the right instrument for the discovery of

Earth-like planets in orbit around other stars. However, it will improve on the Hubble Space Telescope. In order to overcome this difficulty NASA is considering a Space Interferometry Mission.

Its purpose is to place a very large infrared interferometer in an Earth-like orbit around the Sun: the "Terrestrial Planet Finder" [7] (TPF). Its dimension would be something in the range of 75-100 meters. It would have four separate telescopes that will be able to operate in the infrared (IR) part of the electromagnetic spectrum. In order to reduce the light from the central star, it will combine carefully the IR radiation detected by each of the four telescopes.

The TPF will have to be in orbit around the Sun at about a distance of 5 AU. In this case the zodiacal dust cloud [13] can be largely avoided. This dust reduces the energy received from stars to such an extent that, by avoiding it the aperture of each of the four telescopes to be placed along TPF can be reduced from twice the dimension of the Hubble Space Telescope down to only half such an aperture. This implies a very practical point: reducing the costs might make its completion in a much shorter time. We have argued in Chapters 7 and 8 that our solar system may provide evidence of extraterrestrial life in the form of microorganisms. Projects such as the TPF may allow us a glimpse at a much wider sample of worlds.

On the habitability of satellites around Jupiter-like planets

It is no longer evident that the habitability zone, in which planets and their satellites may develop life, need to be found in a limited region near the star. In the case of our solar system, the habitability zone has been taken to lie outside the orbit of Venus (0.7 AU) and beyond the orbit of Mars (1.5 AU). What we do know about Europa suggests that the characteristics needed for sets of extrasolar satellites around Jupiter-like planets to harbor life are (cf., Chapter 8) :
• A non-circular orbit (caused by the gravitational interaction of the remaining satellites of the given planet).
• The vicinity to a sufficiently large planet, so that tidal interaction may be effective in producing hydrothermal vents, favorable for bacterial ecosystems.

We should underline that all that is relevant is the presence of large Jupiter-like planets, not that the Jupiter-like planet be near the companion star. Traditionally the only concept that has led to estimating the boundaries of the zone of habitability has been the reasonable distance of the Earth-like planet from its star.

From these statements we may derive a positive lesson form the point of view of astrobiology: even though Earth-like planets have not yet been detected, and even though giant planets are not evident environments that may support life (although this possibility was discussed in the past), the new extrasolar planets are already indicators of possible environments favorable for the origin of life. This conclusion follows from the generality of the arguments of the formation of Jovian planets from their corresponding subnebulae (cf., Chapter 2). The mechanisms for the formation of satellites suggest that in the star cradles in the Orion nebula [14] satellite formation around Jupiter-like planets should be similar (cf., Fig. 10.2).

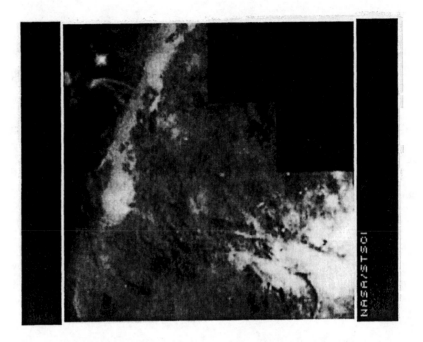

Figure 10.2: The Orion nebula is a gas cloud, in which the younger stars are surrounded by disks of dust and gas, not unlike our own solar system in dimension. (Courtesy of NASA/Hubble Space Telescope.)

11

The search for the evolution of extraterrestrial intelligent behavior

One of the greatest scientific achievements that has marked the Second Millennium is to begin the search for the place of Homo sapiens sapiens in the universe. Two great contributions in this search did occur in the 16th and 19th centuries.

What is our place in the universe?

The present chapter will begin to highlight contributions that have given us a vision of our own place in the universe. We shall postpone further comments of historical nature to Chapter 15; we prefer to continue to sketch in Chapters 13 and 14 the scientific, philosophical and theological questions that serve as a frame to the main questions of astrobiology.

The pioneering work of Nicholas Copernicus revived the heliocentric theory, which had been formulated two-thousand years ago by Aristarchus of Samos. As we shall see in Chapter 15, this major scientific achievement, together with the intuition of Giordano Bruno, who extended the Copernican view to an infinite universe, and Charles Darwin with his theory of evolution, we have finally begun to infer a balanced view of the real position of the Earth in the cosmos, as well as the position of the genus *Homo* within the Earth biota.

On the other hand, what remains to be settled in the new science of astrobiology is to learn what place the Earth biota occupies in the cosmos. In other words, as we have emphasized elsewhere in this text, the distribution of life in the universe remains as the central question to elucidate by further research and astronomical observation. But astrobiology is not only concerned with establishing the presence of life in the universe. An ultimate objective is to know firstly, whether we can detect the signature of intelligent behavior on other worlds; secondly, whether we can communicate with other beings. These are the topics we shall discuss below, and in Chapter 12.

Intelligent behavior is intimately related with the need to communicate

As mentioned in Chapter 1, powerful searches have been implemented since the early 1960s for intelligent radio signals, a question whose discovery would fill the missing gap that all scientific research during the Second Millennium has been unable to answer. Success in such a fundamental program would directly answer the question of whether we are alone in the universe.

The discipline is already almost 40 years old. Edward Mills Purcell was co-discoverer with Doc Ewen in 1951 of the 21-centimetre line, the first spectral line to be discovered by radio astronomy (microwave emission of hydrogen atoms in deep space). The serious systematic "search for extraterrestrial intelligence" (SETI) began in 1958 with the suggestion by Giuseppe Cocconi and Philip Morrison [1] that a search be conducted at radio wavelengths emitted by instruments devised by intelligent beings that may have developed beyond our solar system.

Intelligent behavior amongst evolved creatures seems to be intimately related with the need to communicate amongst themselves. Cocconi and Morrison suggested that a reasonable wavelength for communication might be that of the hydrogen line at twenty-one centimeters. Indeed, hydrogen is the most abundant element in the universe. It may be found in clouds and in interstellar space. When it absorbs some energy, it releases some of it by the emission of radio waves characterized by a wavelength of 21 cm. Due to its ubiquity, the 21 cm line is now used as one of the principal means for exploring the structure of galaxies through radio observations.

Progress in instrumentation

In the almost four decades that this search has been maintained, a great deal of progress in the necessary technology has been achieved [2] . After the first forty years no signal has been detected [3], but the equipment that is currently in use is about 100 million times more powerful than the equipment that was available in 1960. In fact, current searches with Project Phoenix, SERENDIP and BETA each are sampling 200 million frequency channels (cf., Table 11.1). The main enquiry may be formulated as the *Cocconi-Morrison question:* What waves would be most likely to be used by intelligent beings across the enormous interstellar distances that separate, for instance, an Earth-like planet or satellite in *Pegasi 51* [4] from the Solar System? The answer to this question involves several factors, but the SETI 'window' is 1-10 GHz, in other words, this would correspond to wavelengths in the radio range. In this domain radio astronomers have already the equipment to face the problems raised by SETI research.

The Phoenix Project

Phoenix arose form a previous NASA-funded project (the High Resolution Microwave Study, HRMS) [5, 6] . Private funding however was the source that preserved this all-too-important research effort. From 1993 till the end of the decade the original NASA-funded equipment was preserved and much improved (cf., Table 11.1). The funding is managed by an independent corporation headed by Frank Drake at the SETI Institute in Palo Alto, California. The Director of Research for the SETI Institute is Jill Tarter.

In the Phoenix Project, originally directed by Bernard M. Oliver [7], more than one radio telescope is used. The current director is Jill Tarter. In the first phase of the project the main telescope was the 65-meter dish at Parkes, Australia; the complementary telescope measures 30 meters and is located at Mopra, Australia.

The main objective was to observe target stars in the southern hemisphere over frequencies in the range 1-3 GHz. Then Phoenix operated at the NRAO Green Bank in West Virginia coupled with a facility at Woodbury, Georgia. At present the main objective is the northern hemisphere and the Arecibo radio telescope in Puerto Rico is being used.

The search proposed initially to cover about 1000 target stars within 160 light years. The impressive sensitivity that has been achieved is capable of detecting the analogues of strong terrestrial radar signals.

TABLE 11.1: A selection of some SETI Projects [6].

	BETA	*META II*	*SEREN-DIP IV*	*Southern SEREN-DIP*	*Medicina (Italy)*	*Phoenix*
Site	Harvard	Buenos Aires	Arecibo, Puerto Rico	NSW, Australia	Bologna	Several
Antenna diameter (m)	26	30	305	64	32	22-305
Channels (millions)	250x8	8.4	168	4.2x2	4.2x2	28.7x2
Approx. sky coverage (%)	70	50	30	75	75	targeted survey
Spectral resolution (hertz)	0.5	0.05	Down to 0.6	0.6	1.2	Down to 1

The META, BETA and SERENDIP Projects

META is an acronym for *Megachannel ExtraTerrestrial Assay*.   In this project the basic idea is that there are no known processes in stellar dynamics that are capable of producing sharp lines as the one suggested by Cocconi and Morrison. If such a signal is received, then the natural conclusion is that an intelligent source is responsible for its emission [7].

The META project used a 26-meter telescope at Harvard. Over a one-year period in cooperation with a radio telescope in Buenos Aires, Argentina, the complete celestial sphere is scanned for the presumably artificial signals that might be indicative of another intelligence. Over a period of 5 years several candidate signals were detected, but none confirmed.

The BETA project has been developed in order to overcome the limitations of the previous META project (mainly the fact that signals are one-shot events that cannot immediately be reconfirmed). Other initiatives are based in the United States (Berkeley), Italy (Medicina near Bologna), and Australia (at the Parkes Observatory in New South Wales, NSW). We refer the reader to Table 11.1 for some further details of these projects.

SETI   on the Moon

As there is a background level of radio noise from radio, television, airplanes and telephones amongst others, it is natural to search for locations isolated form this impediment to SETI. One reasonable proposal is to move the research to the far side of the Moon in the 100 km diameter Saha crater in the Mare Smythii, as advocated initially by the late Jean Heidmann (1923-2000) from the Observatory of Paris [8]. This French radio astronomer emphasized that great limitations imposed on radio astronomy from the ground, due to the growth of radio frequency interference arising from commercial uses of the electromagnetic spectrum, particularly radio and TV communication.

The position of the Saha crater (102° E, 2^O S) permanently keeps it away form electromagnetic pollution due to activities on Earth (cf., Figure 11.1).   None of the extremely ambitious engineering projects underlying this proposal seems beyond present technological capabilities, such as the construction of a 340 km road linking laboratories in *Mare Smythii* (at the equatorial level and at the Moon's limb) and crater Saha.

There are several reasons why the Saha crater is appropriate for planning the Moon base for future SETI research. From its center the rim has its base depressed under the horizon by half a kilometer.

This feature of the crater is due to the lunar curvature The consequence of this particular aspect of the crater morphology is that from the perspective of an observer at the center of the crater, the three-kilometer high rim would only rise a few degrees from the observer's horizon.

Other remarkable aspects of this crater is that it is very near the lunar equator and also to the near side of the Moon, although it is never exposed to electromagnetic radiation (including radio and TV signals) coming from the Earth.

For evident reasons, it is likely that a base will be established on the near side of the Moon before the radio astronomy equipment may be taken to the far side. It is estimated that the Saha SETI base may be a reality by the second or third decade of this century. It seems possible to start establishing a small automatic radio telescope in the crater, which can be operated form the Earth by means of a radio station located in Mare Smythii [9].

The engineering difficulties involved in linking the observational base in the crater with Mare Smythii (where there could be an Earth-link antenna) are not insurmountable. A regolith pathway only 350 km long would be sufficient for this purpose. It should be remembered that lunar landscapes are very smooth with well-granulated regolith.

Figure 11.1: The Saha crater, an image taken by Lunar orbiter II in 1966 (courtesy of NASA).

The technical progress of SETI

The collaboration between the SETI Institute and the University of California are planning at the time of writing the first major telescope dedicated entirely to SETI. It is referred to as the One Hectare Telescope (1HT). It should be functional within a decade. One of its main features is the ability of improving its size with time and further funds[10]. Indeed, the new telescope will combine the signals from an array of over 500 dishes just 3.5-5 metres across.

By combination of signal form the various components of the telescope, it will be able to make high resolution observations from 100 different directions at any one time. Its range of sensitivity spans the frequency range 300 MHz-10 GHz.

Several major world-wide projects, now underway, will develop designs for cost-effective 1-square kilometer telescopes; the Netherlands and Australia are planning to build these instruments, while a Chinese effort anticipates to construct 30 Arecibo-sized telescopes in the south-west region of the country.

On the difficulties of the SETI project

The eventual success of the SETI project seems inevitable. However, making the first contact with an extraterrestrial civilization has been delayed for some good reasons. They have been mentioned by one of the leading researchers, Paul Horowitz (cf., the BETA project) [11] :

> *The hard part* [of SETI] *is the last step, which is intelligent life in the galaxy transmitting radio waves to us at the wavelength that we are expecting, and at a power level such that we can detect them. That is a lot of ifs.*

From the first neuron to brains

We have divided the subject of astrobiology into three 'books', partly to emphasize two distinct aspects in which the scientific bases are on different footing. Book 1 is supported by the fields of chemical evolution and Darwin's theory of evolution, both of which are time-honored sub-disciplines of astrobiology. In Book 2, we discussed some relevant aspects of biological evolution. Finally, in Book 3, we still lack an underlying theory.

We mentioned earlier, in Book 1, the transition form a simple prokaryotic blueprint of the Archean to the eukaryotic world of the Proterozoic, and Phanerozoic eons. We attempted to show that eukaryogenesis was probably the most transcendental step in the pathway that led from bacteria to Man. Once eukaryogenesis took place on Earth (Chapter 5), the steps leading up to multicellularity (Chapter 6) [12] , namely,

• cell signaling and
• the organization into cooperative assemblies (tissues),

were inevitable from the point of view of evolutionary advantage (cf., "The phenomenon of multicellularity", Chapter 6).

The onset of multicellularity is also due to the considerably larger genomes that were compatible with the eukaryotic blueprint. The densely-packaged chromosomes in the cellular nucleus presented multiple options for opportunistic ways of passing genes to progeny, some of which allowed their carriers to be better adapted to the environment. Mitosis was a more advanced process of cellular division than simple prokaryotic fission. Such variety of options were the raw material for natural selection to improve upon the three billion year old single-cell strategy of life on Earth. The improvement was achieved within 30% of the single-cell 'era'. A full-organism strategy led subsequently to large-brained organisms and intelligence.

The first neuron could not arise at the prokaryotic level of development. A complex pattern of gene expression is required for a functional neuron; consequently, the first stage of a nervous system had to wait for the eukaryotic threshold to be crossed.

The superior strategy of translation of the genetic message of nucleated cells permitted a variety of proteins to be inserted into the cellular membrane [13]. This new stage in evolution provided all the channels necessary for the alteration of the ion concentration inside and outside the cell. Indeed, early in the evolution of the eukaryotes we can conjecture that electrochemical imbalances were already being created by a disparity of electric charges on both sides of the cellular membrane.

The details of the first steps towards multicellularity are not known in detail, but it is evident that in the struggle for survival, evolutionary pressures were going to encourage cells whose genomes codified proteins for cell signaling, as well as proteins that favored the formation of tissues of primitive multicellular organisms.

We have not insisted in Book 1 on the complete understanding of all the intermediate steps that led from organic chemistry to life (still an open problem). In attempting to understand the distribution of life in the universe, we follow a similar strategy. In fact, we do not insist in filling in all the molecular biology details that are involved in the transition form the world of single-celled eukaryotes to multicellular organisms, neurons and brains. (We shall return to this topic in Chapter 12.)

The Drake equation

The key to the central problem of Book 3, namely the definition of the science of the distribution of life in the universe is narrowed down to the problem of eukaryogenesis itself (cf., Chapters 5 and 6). The Drake equation may help us to drive the point home. The word equation need not deter any non-mathematical reader, since as Frank Drake himself pointed out [14], his equation was originally a simplification of an agenda for the points to be considered during the first meeting in 1961 in which SETI was to be discussed. The key parameter in the search for other civilizations is written as f_i, denoting the fraction of life-bearing planets or satellites, where biological evolution produces an intelligent species. An early reference regarding the transition from prokaryotes to eukaryotes in relation with the Drake equation was made by Carl Sagan. His remarks were in the context of a discussion of the SETI projects at a 1971 conference [15]. In order to make Sagan's general comment more specific, we consider the Drake equation, which is explained in Table 11.2.

TABLE 11.2: A simplified form of the Drake equation.

$$N \; = \; k \, f_i \, ,$$

N is the number of civilizations capable of interstellar communication,
k is a constant of proportionality involving several factors that we need not discuss here,
f_i, denotes the fraction of life-bearing planets or satellites, where biological evolution produces an intelligent species

Eukaryogenesis as a factor in the Drake equation

It is useful to point out that the Drake parameter f_i is itself subject to the equation explained in Table 11.3. Our conjecture motivates the search *within our own solar system* for a key factor (f_e) in the distribution of life in the universe, including intelligent life. The extrapolation of the transition to multicellularity into an extraterrestrial environment is suggested by the selective advantage of organisms that go beyond the single-cell stage. Such organisms have the possibility of developing nervous systems and, eventually, brains and intelligence.

If prokaryogenesis is possible in a short geological time frame, there are going to be evolutionary pressures on prokaryotes to evolve, due to symbiosis, horizontal gene transfer and natural selection (cf., Chapter 4).

TABLE 11.3: A simplified form for the equation for the f_i parameter.

$$f_i \quad = \quad k\,f_e \,,$$

f_i, denotes the fraction of life-bearing planets or satellites, where biological evolution produces an intelligent species,
k is a constant of proportionality involving several factors that we need not discuss here,
f_e denotes the fraction of planets or satellites where eukaryogenesis occurs.

In fact, these evolutionary mechanisms are going to provide strong selective advantage to those cells that can improve gene expression by compartmentalization of their genomes. (Larger genomes would be favored, since organisms with such genetic endowment would have better capacity for survival, and hence better ability to pass their genes to their progeny.) Whether the pathway to eukaryogenesis in a Europan-like environment, or elsewhere in the cosmos, has been followed, is clearly still an open question. We already know that f_e is non-vanishing on Earth. What is summarized in Table 11.3 is that, instead of searching for intelligence, the more restricted search for eukaryogenesis is sufficient for understanding one of the main aspects of the postulated existence of extraterrestrial life, the basic assumption of any search for either extraterrestrial intelligent behavior, or the search for extraterrestrial eukaryotes.

Testing the Drake equation in the Solar System

As we have seen in Chapters 7 and 8, it is feasible to search for extraterrestrial microorganisms on Mars and Europa. But we must state clearly the theoretical bases that motivate such a search. To face the problems in the new science of the distribution of life in the universe, we could look for some hints at the problems of biology before Darwin. In Chapter 5 we recalled the significant step that Lyell took in his *Principles of Geology*. Lyell pictured a world constantly and slowly changing; in his view knowledge of the present conditions of climate, volcanic activity and Earth movements, were sufficient for extrapolating back in time, in order to get insights into the same geologic questions in the distant past.

This 'doctrine' inspired Darwin when he successfully attempted to lay the scientific foundations of natural history. In volume 2 of *Principles of Geology*, Lyell accepted simple extrapolations into the past of the information we have in the present; but he did not take the crucial step to assume the same gradualism for living organisms as well. That crucial step was left for Darwin to take. In other words, we had to wait for the theory of evolution before we thought of all life on Earth in terms of a single tree of life; species change gradually (by the mechanism of natural selection) and are interconnected by the tree of life (the theory of common descent). Our generation has a similar undertaking, namely to provide a theoretical framework for the distribution of life in the universe.

Today, thanks to molecular biology, we have isolated eukaryogenesis as the key phenomenon in the pathway that leads form bacteria to Man, intelligence and civilization. This insight, however, is restricted to the only tree of life that is known to us. The SETI program, or what we might call the more restricted program for the search for extraterrestrial eukaryotes attempts to make some progress in the missing theory of the distribution of life in the universe. Indeed, SETI directly searches for traces of intelligence of the same level of evolution that we have reached, or even beyond. The reasonable observation of SETI researchers is that, given the antiquity of the universe, there may be planets or satellites older than 5 billion years, still in orbit around a main sequence star that is not near its supernova stage. The advantage of focussing on eukaryotes, as emphasized in this chapter, is that we can proceed to test it in solar system exploration.

To embrace a guiding line, we assume as a working hypothesis that evolution of life in the universe can be explained only in terms of evolutionary forces that we experience today in our local environment. In other words, if other intelligence has developed on extrasolar planets, it must have gone through eukaryogenesis. More precisely, we have conjectured that:

Life is not only a natural consequence of the laws of physics and chemistry, but once the living process has started, then the cellular plans, or blueprints, are also of universal validity: The simplest cellular blueprint (prokaryotic) will lead to a more complex cellular blueprint (eukaryotic). Eukaryogenesis will occur inexorably because of evolutionary pressures, driven by environmental changes in planets, or satellites, where conditions may be similar to the terrestrial ones.

Beyond geocentrism, anthropocentrism and 'biogeocentrism'

Although there are still many questions to be answered, as we saw in Chapter 8, at present it seems possible (although not an easy matter) to penetrate the oceans of the iced Galilean satellites, provided it is confirmed that there are submerged oceans of liquid water. We can conceive experiments addressed specifically to the question of the search for extraterrestrial eukaryotes, although clearly many such experiments could be formulated. To sum up, verifying that f_e (cf., Table 11.3) is non-vanishing in one of the Galilean satellites goes a long way towards making contact with the initial steps towards other intelligence. It also would lay the foundation for the theory for the distribution of life in the universe.

This chapter has summarized the efforts that science has made to free itself form the special way we see ourselves. The preliminary steps taken by science, philosophy and natural theology assumed that humans had a privileged position in the cosmos. With Copernicus geocentricism was abandoned [16]. Natural theology has incorporated this view, after an initial opposition to Galileo's inexorable conclusions (Chapter 8). After Darwin's theory of evolution was formulated, anthropocentrism was abandoned, although more than a century of research was needed, culminating with the solid genetic bases of neo-Darwinism. Natural theology is also beginning to incorporate this second step into its discussions (cf., Chapter 13). We have called, for lack of a better word in the English language, 'biogeocentrism' the concept that life is exclusive to the Earth. The underlying difficulty for many scientists arises from a confrontation between the physical and biological sciences.

Firstly, everyone agrees that Newton's theory of gravitation can be extrapolated without any difficulty throughout the universe, except for the small corrections required by Einstein's theory of relativity (cf., Chapter 1). One example was already discussed in Chapter 10 regarding the orbits of Jupiter-like planets. Secondly, the case of extrapolating the theory of biological evolution throughout the cosmos requires more care and is still an open problem. Arguments against the hypothesis of biogeocentrism can now be formulated thanks to progress in our understanding of Darwinian evolution. The role of randomness has to be qualified since Darwin's time.

Although chance is implicit in *The origin of Species* and Monod captured the essence of Darwinism with the suggestive contrast between chance and the necessary filtering of natural selection, we have also seen that molecular biology greatly constrains chance. Convergence will be a further factor to take into account, as we shall discuss in the next chapter. To sum up, Darwinian contingency is constrained and evolution tends to converge on similar solutions when natural selection acts on similar environments. Cosmochemistry and planetary science suggest that the environments where life can originate are limited. We already are gathering information on a significant number of Jupiter-like planets around stars arising from protonebulae that are likely to grant them an array of satellites. In the outer Solar System this can be confirmed. Each of our giant planets has a large number of satellites.

Such is the case in the outer Solar System. Factors giving rise to atmospheres in Jovian satellites are known. Titan, on the other hand, has an atmosphere produced by outgassing, combined with the seeding of volatiles by comets carrying a fraction of water-ice. Evidence is leaning in favor of the existence of Jovian planets with masses larger than Jupiter; hence tidal heating responsible for Io's volcanic eruptions, could be even more efficient in other solar systems. On Europa it is not yet clear whether tidal heating may produce hydrothermal vents capable of giving rise to life. The case of satellites orbiting around large Jupiter-like planets will have to be investigated in the future.

To sum up, natural selection will be working on a finite number of similar environments. According to cosmochemistry, similar chemical elements will enter into pathways of chemical evolution. We have also learnt that there are no laws in chemical evolution that are specific to the Earth; it is reasonable to hypothesize that biological evolution will follow chemical evolution. Today Darwinism cannot be seen as a simple dichotomy between chance and necessity. On the contrary, we should be well aware that convergent evolution and constrained chance are fundamental ingredients of Darwin's theory. Thus, in environments similar to Earth, we may expect analogous pathways leading from chemical evolution to the origin and evolution of life and, eventually, to the evolution of intelligent behavior on other worlds.

12

Is the evolution of intelligent behavior universal?

Throughout the book we have emphasized one possibility for evolution of life in the universe: the universality of the eukaryotic cell leading up to not a single tree of life (phylogenetic tree), but rather to a "forest of life". This is in sharp distinction with the view that ours is a 'rare Earth' (cf., Preface). Clearly there are other alternatives, but only experiments carried out in appropriate solar system missions will give us the correct answer. We call the attention of the reader to the fact that experiments designed to select between the alternative pathways of evolution are possible. (It should be underlined though that such experiments are both costly and slow in their implementation.)

Convergence and contingency in evolutionary biology

To begin with, some aspects of evolutionary behavior in animals are worth mentioning: Communication is a general social activity within species that have significant differences in brain organization (ants and humans); in ants communication has evolved by the emission of pheromones singly or in combination, and in various amounts, to say to other ants in effect, for instance: *danger,* or *follow me* [1]. In humans communication has ranged from drum beating in primitive populations to extensive use of the electromagnetic spectrum, as radio or TV signals (during the last century). For the first time to the best of my knowledge, in a recent publication Frank Drake [2] has yet added a fresh angle to our insights into the possible existence of life elsewhere in the Cosmos. He exposes in a conjecture that we shall refer to as the *Drake conjecture*, (the Drake conjecture is being formulated almost exactly 40 years after the formulation of the by now famous Drake Equation, cf., Chapter 11, Table 11.2): In simple terms the Drake conjecture is that there is a possibility that after an easy period of detection of an extraterrestrial civilization, progress of a given group of beings elsewhere, may be followed by a more silent period, because of the improvement of the means of transmitting information.

For instance, the current use of the internet has created a definite trend towards diminishing the use of radio and TV signals, which after all are the indicators of the presence of other civilizations that may have acquired a technology similar to that of humans throughout most of the 20th century, prior to the massive onset of the internet and other means of transmitting digital information.

On the other hand, in this context it is worth remembering that cetaceans and primates, in spite of significant brain morphology, have a certain degree of *behavioral convergence*, such as a prolonged period of juvenile dependency, long life span, cooperative hunting, and reciprocal altruism. We shall use the pragmatic definition of intelligent behavior in those taxons that share such behavioral convergence with humans.

We shall now argue that terrestrial biota teaches us that multicellularity is inexorably linked with neurogenesis, which, in turn, is linked with the process of synaptogenesis, and finally with the formation of cerebral ganglions at the lowest levels of a given tree of life. Once protobrains are formed, brain evolution will be constrained by few mechanisms, prominent amongst these is brain organization across species in terms of, for instance, modules. It should be remarked, in addition, that critical clues have already been provided in developmental neurobiology in terms of molecular genetics, regarding the remarkable increase in brain size per body weight in humans (a concept that will be made accurate below). It concerns insights in terms of *homeotic genes,* whose expression pattern correlates with the repetition of similarly organized segments along the body.

Hence, we may consider brain evolution to be independent, to a certain extent, of historical contingency. Whether nature in an extraterrestrial context steers a predictable course, is clearly still an open question, but some hints from the basic laws of biology militate in its favor: These laws are natural selection and the existence of a common ancestor (a 'cenancestor'). All the Earth biota can be interpreted as evidence in favor of the fact that, to a large extent, evolution is predictable and not contingent. The underlying question concerns the relative roles of adaptation, chance and history. This topic is subject to experimental tests.

We shall assume that natural selection seems to be powerful enough to shape terrestrial organisms to similar ends, independent of historical contingency. In an extraterrestrial environment it could be argued that the evolutionary steps that led to human beings would probably never repeat themselves; but that is hardly the relevant point:

> the role of contingency in evolution has little bearing
> on the emergence of a particular biological property.

Besides, it can be said in stronger terms that essentially evolutionary convergence can be viewed as a 're-run of the tape of evolution', with end results that are broadly predictable. (This question will be taken up again in Chapter 13, cf., "Can our intelligence be repeated elsewhere?.)

The inevitability of the emergence of particular biological properties is a phenomenon that has been recognized by students of evolution for a long time. It is referred to as 'evolutionary convergence'. This may be illustrated with examples taken from malacology, and ornithology, as we have done in Chapter 13.

To discuss whether intelligent behavior is a logical consequence of the evolutionary process, we will recall the definition of astrobiology given in the Introduction: it is the subject that studies the origin, evolution, distribution and destiny of life in the universe. For clarifying our ideas, we may consider briefly, the overlap of these sub disciplines with neuroscience:

"THE ORIGIN OF LIFE IN THE UNIVERSE" is a well studied discipline. The main lesson that we may derive from what we have learnt in the last 80 years, when the subject started with the experiments of Alexander Oparin, is that life's origin occurred on the early Earth, almost as soon as it could possibly do so. We shall return to this later, when we shall discuss the possibility that in just 200 million years after the formation of the Earth the conditions were suitable for the origin of life.

This geologic discovery of early biofriendly environments should be seen together with the earliest evidence of microfossils, which suggests that just one billion years after the origin of the Earth, there were microorganisms of considerable evolution, such as cyanobacteria colonies, which are known as stromatolites.

"EVOLUTION OF LIFE IN THE UNIVERSE" is precisely within the scope of all biologists. For although life is still only known in one solar system of one galaxy - the Milky Way - we know a considerable amount, particularly in the topic of the evolution of intelligent behavior.

"DISTRIBUTION OF LIFE IN THE UNIVERSE" brings us closer to the subject of the possible universality of the evolution of intelligent behavior. For over 4 decades the SETI Institute under the direction, firstly of Drake, and presently of Jill Tarter, have made systematic progress in radio astronomy searching at various frequencies in the electromagnetic spectrum for signals from intelligent social groups of creatures that communicate amongst themselves. The technological progress achieved by the SETI researchers has been considerable, at present billions of channels can be processed simultaneously. There are no reproducible intelligent signals so far.

Finally, "DESTINY OF LIFE IN THE UNIVERSE" does not overlap at all with neuroscience, it rather opens the door to interaction with other sectors of culture that normally do not interact directly with science itself. I have in mind philosophy and theology.

The main lesson we derive from the combined evidence of early geologic evolution of our planet, together with the earliest reliable micro-fossils, is that given a planet such as the Earth, the evolution of prokaryotic cells is intimately bound with the earliest stages of planetary evolution. We hope to have placed the topic 'evolution of intelligent behavior' within the wider frame of astrobiology. What I have to say draws a powerful lesson from the origin of life on Earth and lies somewhere between the evolution and distribution of life in the universe.

Biological evolution on other worlds

Let us discuss the rationale for the search for extraterrestrial signals of the evidence of intelligent behavior:

We think that a central nervous system is bound to evolve, since in the conventional phylogenetic tree, already at the level of the diploblastic animals (cnidarians, with the familiar example of jellyfish), there has been electrophysiological research demonstrating that such simple animals do have nervous nets. In Table 12.1, we should emphasize that the simpler multicellular arrangements, namely sponges have electrophysiological responses; but no primitive nervous systems have been demonstrated, for instance, similar to the nervous nets of the cnidarians. To sum up, almost as soon as some coordinated electrophysiological responses are possible in multicellular organisms, they have been demonstrated to exist.

We feel that the brain is a sure thing to evolve. In the pylogenetic tree, the first appearance of a cerebral ganglion occurs very early, for instance in annelids (ancestors of common worms). Once again, as soon as the diploblastic/triploblastic barrier has been crossed, cerebral ganglions appear.

The more difficult case to study is that of the first steps towards intelligent behavior and complex language. This is what the SETI project assumes to be occurring on other worlds. However, we already know that in humans the origin of language is probably a consequence of natural selection. There is a long debate on this issue, which started with Chomsky and was taken up by Pinker (cf., Chapter 6).

Without entering into the details of this argument, it seems reasonable to assume that natural selection may favor the appearance of language, once a sufficiently complex brain has appeared in a given phylogenetic tree.

Table 12.1: Some physiological responses at the lowest level of the phylogenetic tree[3].

Organism	Physiological responses
Paramecium (protozoa)	Calcium channels involved in the protozoan movements
Rhabdocalyptus dawsoni (sponge)	Ca- and Na-dependent channels
Aglantha digitale (jellyfish; cnidarian)	Action potentials have been characterized (nervous nets)

The case of humans illustrates that the evolution of complex intelligent behavior, including the evolution of communication through language, rather than chemical signals (pheromones), has been favored by the emergence of mutually supportive social groups of creatures, at the later stages in the evolution of a complex brain.

Evolution of intelligent behavior in an aquatic medium

On Earth primates and dolphins present us with examples, which demonstrate that sharp increments in brain size can probably occur in different environments: terrestrial and aquatic. The dramatic increase in the last 2 million years (Myr) in relative brain size, compared to body weight (a concept that will be defined accurately below) has been considered as a hallmark of humans, and their special place in the universe.

There may be plausible mechanisms in terms of which the rapid increase of human brain size may be rationalized. Preliminary arguments in terms of *homeo*genes and gene duplications have been put forward. But independent of the molecular mechanisms that may be considered to be at the bases of hominid evolution, the anthropocentric heritage still within science gives undue attention to the last 2 Myr as a frame of reference for further discussions.

This may be inappropriate, since there is some evidence, which extends the favorable conditions for the origin of life on Earth back to 4,400 Myr before the present. In other words, looking at the ascent of humans above other species, as an example of an unlikely, or a unique event, is at least misleading. Our brain has evolved remarkably only during the 0.5% of the history of life on Earth. We should look at other groups beyond the anthropoids, to which we belong, to verify whether within a brief period of geologic time, animals with different brain morphology, have gone through other episodes of a many-fold increase in brain size per body weight, above the average of similar sized animals. We shall return to the example of cetaceans that had a dramatic 9-fold increase in brain size per body weight over the relatively short period of the past 45 to 50 Myrs. In fact, it is pertinent to dwell on this point a little longer. One of the correlates of intelligence is relative brain size. Across the barrier of species, we can assign a quantitative parameter, the encephlization quotient (EQ) [4]. The EQ parameter reaches a value of 7 in *Homo sapiens* and 4.5 in some cetaceans; the parameter has lower values in non-human primates.

Encephalization is defined as the increase in brain size over and above that expected on the basis of body size. EQ is determined relative to a sample of species, so that its absolute value expresses the level of encephalization, relative to the rest of the species in the sample. This definition avoids the 'chihuahua conundrum': These small dogs are much more encephalized than a bull-dog, but are in no significant way smarter. This is why the definition of encephalization is with respect to the rest of the species.

The value for humans is to be interpreted as expressing that humans have, on average, a brain seven times larger than that of a mammal of similar body size. The point that is really relevant to the question of the evolution of intelligent behavior in a given species, is that in the past 1-2 million years, the human EQ and that of cetaceans were equivalent [5].

Experimental tests for the evolution of microorganisms elsewhere in our solar system

This remark is pertinent, since it is only after the surveys of Voyager in the 1970s and of the Galileo Mission in the 1990s, that astrobiology began to take seriously the possibility of biological evolution in aquatic biotopes, other than terrestrial ones. Experimental tests on the evolution of microorganisms are of considerable interest in the oceans of the Solar System, or in isolated reservoirs on Mars. We proceed to list some key questions that remain to be clarified in the crucial evolution experiments.

We mention here just a few of the outstanding difficulties: When we argue that evolution of a nervous system is an advantage for sensing the environment, we are only partly right. This is only true if there is need for motion or, more importantly, active predation. The truth of this statement is clear if one examines the other two multicellular kingdoms, plants and fungi, in which motion (although it is present) is not evident.

David Attenborough [6] has emphasized that swiftly moving creatures such as ourselves, tend to regard plants as immobile organisms, rooted to the ground. But the production of new individuals leads to the extension of the domain of their species. Motion in plants is normally invisible to the naked eye. However, the fact remains that we have illustrated some evidence that the first neurons, as well as primitive nervous systems in the phylogenetic tree of life, can be traced back to the simplest multicellular organisms. Such organisms evolved much earlier in time than the subsequent first appearance of plants and fungi. Granted that the argument of active predation excludes plants and fungi, Earth biology has shown that the first steps towards a nervous system predate the evolution of plants. Several phyla, such as Chordates, Mollusks and Arthropods can be traced back to the early Cambrian, earlier than the first appearance of the divisions of the Plant Kingdom. There are a few unifying threads in the process of cellular complexification: association, differentiation, patterning and reproduction - they have been discussed already. It is reasonable to assume that cells first got together as a result of chance mutations, which favored the multicellular association. In turn, they stayed together because they reproduced more successfully as a group, rather than as single cells. Today we can appreciate that slime molds can be seen as a model of how the first steps in multicellularity did occur. These organisms are eukaryotic and heterotrophic.

These unicellular microorganisms segregate a chemical - cyclic adenosine monophosphate (cAMP), which leads to aggregation into a single macro-organism. Indeed, each unicellular eukaryote, on contact with cAMP, expresses new surface molecules with 'lock and key' possibilities that after they randomly come into contact, they remain locked to each other [7]. There are additional processes for holding the eukaryotes together: this occurs directly by the expression of surface cell-adhesion molecules, which play a role of holding the cells together and also of linking to extracellular 'scaffoldings' built-up by means of substrate adhesion molecules.

Underlying this ability of the eukaryotes to form efficient and functional communities, there is a major role played by sexual reproduction, rather than the simpler reproduction processes of the prokaryotes (for instance by fission). But in the present context we shall not elaborate on this key aspect of evolution and diversification of the eukaryotes, as part of multicellular organisms. The earliest multicellular group that evolved simple nervous systems are cnidarians and ctenophores. As summarized in Table 12.2, extensive work has been done with the jellyfish *Aglantha digitale*, in which action potentials have been characterized. Hence, the workings of the primitive nervous system (called technically a 'nerve net') of Cnidaria is understood, but they are essentially diploblastic organisms (ie., during development they just have endoderm and ectoderm, what is missing is the 'mesoderm').

It is in the next step in the evolution of primitive animals, we should consider the triploblastic animals. They are animals that have a well-developed mesoderm. It is at this level in which primitive brains are first seen to evolve. In flat worms, there is one example, the *Notoplana acticola*, which has a primitive brain (a cerebral ganglion), receiving inputs form sensory organs and delivering outputs to muscles, via nerve filaments. This has already been summarized in Table 12.2. The evolution of higher animals occurred explosively during the Cambrian, over 500 million years ago.

Soon enough after the emergence of the simplest triploblastic body plans, multiple phyla appeared in the Cambrian, which successfully persevered through subsequent geologic periods, right up to the present: we would like to underline particularly phyla in a major group of animals: Chordates, Mollusks and Arthropods. In all of them we find nervous systems with the capacity to support sensorial discrimination, learning, social behavior and communication. The property of communication through the spoken language had to await the advent of the vertebrates to reach the extraordinary possibilities developed by humans, that are being searched elsewhere in the universe, by means of the SETI project.

Table 12.2: The emergence of cerebral ganglions at the lowest level of the phylogenetic tree [3].

Organism	Cerebral ganglion
Notoplana acticola (flat worms; Platyhelminthes)	Inputs are received form sensory organs and outputs are delivered to muscles, via nerve filaments.
Ascaris lumbricoides (round worms; nematode)	Signals are received from sensory organs and output signals are sent to muscles; peptidergic components are present.
Caenorhabditis elegans (round worms; nematode)	Signals are received from sensory organs and output signals are sent to muscles; peptidergic components are present.

In the above-mentioned phyla of the Animal Kingdom (Domain Eucarya), their nervous systems are considered well-developed, according to their capacity for giving support to sensorial discrimination, learning, communication and social behavior.

The lesson that we are continually learning from astrobiology is always one of unveiling environments that are not too dissimilar to the early Earth, when the evolutionary process of our phylogenetic tree started. Some of these environments are nearby inside our own solar system, namely on Europa and Mars. But many other environments are also coming to our attention. Indeed, no less than 50 planets are now known to circle around nearby stars. They are all large planets, comparable to our own Jupiter. In fact, the question

Are there constraints on biological evolution in solar systems?

is not hopeless. In one or two decades, we will be able to observe earth-sized planets in nearby stars.

Which precursors of evolution of intelligent behavior should we search for?

The question arises as to what properties should we test for the first microorganisms that are currently being searched inside the solar system - The European Space Agency (ESA) will attempt to search for microorganisms in just two years time, with the Mars Express mission, using the Beagle 2 lander, largely the initiative of English scientists.

Some experiments should be discussed to test whether evolution of intelligent behavior is a logical consequence of the emergence of a eukaryotic cell, which is the minimum cellular plan that will allow a neuron to emerge.

These questions are not beyond the range of exploration of the Solar System. Large brains can only occur in multicellular eukaryotic organisms, since the eukaryotic cell will allow genomes with a sufficiently large number of genes to code for the necessarily complex set of ion channels that are needed for a functional nervous system. As mentioned above, in simple eukaryotes, such as *Paramecium,* electrophysiological research has demonstrated that simple microorganisms do have mechanisms for sensing and responding to a variety of stimuli by means of a set of ion channels. The basic question lies on the emergence of ion channels in microorganisms.

Work done with mutants deficient in distinct channels has led to the identification of channels involved in *Paramecium* movements. Subsequently, in the simplest nervous systems of cnidarians, once again, ultrastructural studies have identified ion channels. In an extraterrestrial setting a very preliminary evolutionary experiment that could be discussed would be if, in a given biotope, the microorganisms are eukaryotes that have already developed ion channels. If that stage has been reached then, it is possible that over geologic time, such a biotope would lead to multicellularity and, eventually, to brains and intelligent behavior.

What we have discussed in this chapter is a set of possible evolutionary experiments, whose aim would be to test whether the first steps towards the evolution of intelligent behavior have taken place on other worlds.

Part III

CULTURAL FOUNDATIONS
FOR THE DISCUSSION OF
THE DESTINY OF LIFE IN THE UNIVERSE

13

Deeper implications of the search for extraterrestrial life

An answer to the fundamental question of the relation man/universe, requires a broad cultural discussion. It was characteristic of the Enlightenment, the movement of ideas current during the 18th century- to distrust tradition in cultural matters. Truth was to be approached through reason. At the end of that period Auguste Comte (1798-1857) founded a movement which advocated that intellectual activities should be confined to observable facts. The reason why this movement was called "positivism" is that observable facts were called 'positive' by Comte. This point of view was developed much later by a group of philosophers working in Vienna at the beginning of the 20th century. They were known as the "Vienna Circle". They maintained that scientific knowledge is the only kind of factual knowledge.

Positivism and the Vienna Circle

The Vienna Circle maintained that all traditional doctrines are to be rejected as meaningless. They went beyond positivism in maintaining that the ultimate basis of all knowledge rests on experiment. Since they were also considering the unification of science and were using mathematical logic in their formulation, the Viennese version of extreme positivism came to be known as 'logical positivism'[1]. Although some scientists have adopted this philosophy, either consciously or unconsciously, the fact remains that modern science begins with Galileo, who initiated the tradition of formulating theories based on observation and experiments. No underlying philosophy was adopted then, or need to be adopted now, beyond the dialogue theory/experiment.

159

On the other hand, there is a large number of issues that science cannot handle, or even formulate. In his *History of Western Philosophy* Bertrand Russell makes this point in definite terms [1]; he emphasizes that practically all questions of most interest in theoretical matters are of a nature that even science cannot answer.

Positivism avoided all considerations of ultimate issues, including those of metaphysics and religion. However, as anticipated by Russell, the reduction of all knowledge to science is a matter that debate has not yet settled.

Issues of first causes and ultimate ends are precisely topics relevant to the subject matter of this book. To 19th-century scientists, the problems of the origin and distribution of life in the universe were issues that were to be excluded from the scientific discourse.

In the three Books into which the subject of astrobiology has been divided, we have attempted to show that these problems are approachable by scientific methods. These subjects have a consistent history of valuable efforts by some of the best scientific minds of the 20th century. The long list of such scientists began with Alexander Oparin, John Haldane and included many others, some of which are listed in the name-index at the end of the present work.

In view of the progress that has been achieved, extreme aspects of first causes and ultimate ends are naturally inserted in the science of astrobiology; yet, neither of the two problems (origins and distribution of life) is solidly set on scientific bases: it has been impossible to synthesize a living organism so far, and no signal from an extraterrestrial civilization has yet made contact with the highly sensitive radio telescopes that have been discussed in Chapter 11.

In view of this unsettled state of affairs, it seems unavoidable that a reasonable collective approach to the deepest questions in astrobiology should be encouraged by all sectors of culture.

Regarding the limits of science, inspired by Russell probably a pertinent question is whether man is really what he seems to the astronomer, a small carbon and water organism restricted on a small planet of no particular importance.

To address the Russellian question, we must first decide on the place of humankind amongst the Earth biota.

Position of humans in the totality of all earthly species.

From the perspective of biology, human beings represent only a single species among four thousand mammals. Yet, this is a small number when compared with the 30 million species that are expected to constitute the whole of the Earth biota. One aspect of this bewildering abundance of species of which humans belong to only one, has led to what has been called a metaphor, which expresses the belief that if the history of evolution were to be repeated, for instance, in an extra-solar terrestrial planet or satellite, then such a world would possibly have multiple forms of life but, according to the metaphor, amongst those species, human-like beings would not be included [2]:

Today, a few years after the discovery of extra-solar planets (cf., Chapter 11), this metaphor of 'rewinding the clock of evolution' is no longer a mere example of arm-chair speculation, but can in fact be subject to observation. Indeed, within the foreseeable future mankind will be provided with a new window, through which some of the questions concerning the existence of extraterrestrial life may at last find an answer. The DARWIN project, currently at the planning stage, will be the European Space

Agency (ESA)'s first attempt to search for optical images of other Earthlike planets. The physical properties of our twin planets will be available through spectroscopy in only a few years from the time of publication of this book.

Our main concern is not the origin and evolution of our own species. Our main concern is rather the likelihood that the main attributes of man would rise again, if the history of evolution starts all over again elsewhere, not in a hypothetical Earth that would be miraculously reconstructed. We are mainly concerned with the repetition of biological evolution in an extrasolar planet, or satellite, that may have had all the environmental conditions appropriate for life. The main question is whether the attributes of man are repeatable. Such attributes are, for example, a large brain and consciousness. These features of man evolved from lower primates over the last 5 to 6 million years. This is a short period of time compared to the evolution of other organisms, such as mollusks, which have survived since the Lower Cambrian. Indeed, their first appearance occurred 500 million years ago, about 100 times earlier than the first appearance of man.

Can our intelligence be repeated elsewhere?

In an extraterrestrial environment the evolutionary steps that led to human beings would probably never repeat themselves. However, the possibility remains that a *human level* of intelligence may be favored when natural selection is taken into account. This is independent of the particular details of the phylogenetic tree that may lead to an intelligent (non human) organism. In agreement with Conway-Morris *(The Crucible of Creation)*, for the emergence of any given biological property, the role of contingency may have little relevance in evolution [3]. Instead, the role of convergent evolution should be kept in mind. In the present context of astrobiology the answer to this open question, has to wait until we have access to a second evolutionary line, as a result of either the exploration of the Solar System, or by having evidence of intelligent behavior in other worlds, by means of research in bioastronomy.

I would like to illustrate the inevitability of the emergence of particular biological properties with examples of convergent evolution taken from life on Earth:
• A resemblance amongst shells of distant species of snails has been observed even though the snails in question belong to different taxons (families will be shown in brackets). Examples are known from the Pacific region (Camaenidae), from the New World (Helminthoglyptidae) and from Europe (Helicidae) [4]. In spite of having significant differences in anatomy, these animals have tended to resemble each other in a particular biological property, namely, their external calcareous shell.
• Swallows are included in a group which may give rise to confusion, particularly with respect to swifts. However, swallows are passerines, belonging to the largest order of birds (Passeriformes) [4], a taxon with over 5,000 species. A smaller order, unrelated to the passerines, is constituted by swifts and hummingbirds (Apodiformes). As we have just shown between parenthesis, taxonomically swallows and swifts are members of two *separate* orders, since they differ considerably in anatomy. Their similarities are, once again, the result of convergent evolution in different taxons, the members of which have become adapted to the same life styles in ecosystems that are similar for both species.

In the light of these examples, it is evident that the question of whether our intelligence is unrepeatable goes beyond biological and geological factors. The question is rather one in the realm of the space sciences, in which the radio astronomers may provide an answer, as we have already discussed in detail in Chapters 11 and 12 [5].

In fact, whether we are alone in the cosmos concerns astrometry measurements for the search for extrasolar planets. (Astrometry is the subject that concerns itself with the the observation of position and time variation of celestial bodies.) This activity has led to the current revolutionary view that planets of our solar system are not unique environments that may be conducive to the origin and evolution of life in the universe (Chapter 10) [6].

The presence of a dozen planets in the cosmic neighborhood of the Sun, does not answer positively the question whether we are alone in the universe, but extrasolar planets increase the possible sites where life, given the right ingredients, may evolve. Progress in observational astronomy in the foreseeable future with new projects will begin to provide some answers, such as the possible presence of water and oxygen.

Other constraints imposed on our view of life

A separate question, much closer to our capability to perform practical experiments, concerns the search for microbial life in our own solar system. So, we believe it is appropriate to shift our attention away from 'attempting a full and coherent account of the phenomenon of man' [7]. Instead we should focus our attention at the level of a single cell.

Indeed, we feel that the progress of molecular biology forces upon us a search for a full and coherent account of eukaryogenesis, the first transcendental transition in terrestrial evolution at the cellular level which led to intelligence.

The cosmic search for extraterrestrial intelligence ought, in our opinion, begin with a single step, namely, the search for the first cellular transition on the pathway to multicellularity, and inevitably to brains (due to their selective advantage). This emphasis on the eukaryotic cell as a 'cosmic imperative' has been referred to as the phenomenon of the eukaryotic cell [8,9]. The task of understanding the origins of the eukaryotic cell is not easy [10]. But let me at least dwell on clarifying the terms being used.

Is evolution more than a hypothesis?

We have already reviewed some arguments that suggest that the problem of the position of man in the cosmos critically depends on the evolution of microorganisms up to the level in which eukaryogenesis occurred.

This forces upon us the question of the position of the eukaryotic cell in the cosmos, as the main focus of our attention. I feel that such a radical break with the past has some implications in our understanding of the origin and destiny of man. Nevertheless, none of these arguments lie outside the scope of the question raised in the Papal Message to the Pontifical Academy of Sciences [11]. It is suggested that new knowledge has led to the recognition that the theory of evolution is no longer a mere hypothesis.

In spite of this important step, the acceptance of evolution has not led to a consensus amongst scientists, either on its mechanism, or on its implications. Nevertheless, in spite of this shortcoming, we shall base our subsequent arguments on Darwin's theory of evolution. This leads us to a discussion of the implications that such a search might imply for the dialogue between science and natural theology. We should underline what is really suggested by our current - still uncertain - understanding of the role of evolution in starting a new process, whose end product is complex life: what is suggested is that the broad field of faith and reason - religion and science - should be a fruitful dialogue, rather than a debate.

Constraints on life imposed by philosophy and theology

Constraints imposed by philosophy and theology on our view of life mostly favors a special place of man in the universe. In the case of philosophy there is a continued quest for the impact of technological progress on the future of mankind. We have already encountered the perspective of cultural evolution on the breakdown of a straight coupling between chance and necessity, namely, the continued accumulation of mutations that may favor the adaptation to changing environmental conditions.

A separate question concerns the changes in our theological outlook that may follow the incorporation of knowledge of the place of earthly biota in the cosmos. Would there be problems in the traditional Judeo-Christian-Muslim view of Deity as being confined with the affairs of man? Some scientists have argued that the Deity is omnipotent and can be concerned with the affairs of as many intelligent species as there may exist in the universe.

Yet the question remains whether the original image of God as portrayed in the Scriptures would be acceptable if SETI, or a more restricted search in our own solar system, were to confront us with parallel evolution in other worlds [12].

What is specific to a human being?

A first contact with extraterrestrial life would confront us with new problems to be solved in biology. For example, a more extensive view of taxonomy would be needed. We would have to learn to classify new organisms. This would be within the domain of scientific enquiry.

A different problem, beyond the limits of science, would concern the subject of Divine Action. The monotheistic tradition may be traced back to the New Kingdom of ancient Egypt (1379-1362 BC).

In ancient Israel, on the other hand, writing flourished during the kingdoms of David and Solomon. This progress led, as stated in Chapter 1, to the writing of the Bible, especially from Genesis to Deuteronomy, during the last millennium BC. New Kingdom texts (namely those corresponding to the 18th and 20th dynasties), were written several centuries beforehand. However, we should keep in mind that the revolution of pharaoh Akhenaten (Amenophis IV) was short-lived, in fact, it faded away soon after his death.

Although it is debatable whether the system initiated during the 18th Dynasty may be called monotheism, in *"The Hymn to the Aton"*, close parallels to the verses of Psalm 104 have been pointed out, not only in words, but also in thought and sequence, anticipating this part of the Bible by several centuries [13].

The interested reader who makes the effort to look up the original hymn in the ancient Egyptian text, will find similar references to the creation of the Earth; but he will find more striking still the reference to the creation of the land animals, as well as the creation of birds.

Although it is possible that there may not have been a direct influence between the two texts, they illustrate with specific examples the profound interest that ancient cultures have had on the very questions that astrobiology is attempting to discuss today, but with the difference that now the discussion can be carried out strictly on scientific ground.

A close reading of the Egyptian text would point to the notion of God's acting in the World. This is repeated in the quotation from the Bible.

Closer to our own experience, in the Judeo-Christian-Muslim tradition the ancient concept of Divine Action and its implication in natural theology has been extensively reviewed in the literature [14, 15].

In parallel to the Christian tradition, a position that has been discussed since the Enlightenment regards the confrontation between science and religion. It maintains that God acts only in the beginning, creating the universe and the laws of nature. This thesis is called deism, usually taken to imply that God leaves universal evolution to its own laws, without intervening, once the process of creation has taken place.

On the other hand, within the Christian tradition an approach towards integration has been advocated in the relationship between science and theology. It concerns the problem of biological evolution. Like Darwin, John Paul II, while referring to the living world, for good reasons has not put the main emphasis on first causes and ultimate ends. At the beginning of this chapter we emphasized that for the first time within science there is a branch, namely astrobiology, which makes first causes and ultimate ends its own subject matter.

We identify as a first cause the origin of life in the universe; the distribution of life in the universe may be identified as an ultimate end. With respect to human beings there is much ground to cover yet in the road of convergence between science and religion.

Once again, the Papal Message refers to remaining points still to be discussed. One of the main points raised is firstly that with man we may find ourselves in the presence of an ontological discontinuity.

But this in turn may imply the posing of such ontological discontinuity run counter to that physical continuity that characterizes the main thread of research into evolution in the field of physics and chemistry

Secondly, this argument continues with the consideration of the method used in the various branches of knowledge, which makes it possible to reconcile two points of view which normally would seem irreconcilable. Observational sciences - a main feature of all the basic sciences - describe and measure the multiple manifestations of life with increasing precision.

Finally, this message points out that the moment of transition to the spiritual cannot be the object of experimental science, in spite of all the progress in other aspects of what is specific to the human being.

We have already touched upon the concept of evolutionary convergence. We do not consider the 'rewinding of the evolutionary clock' as a thought-experiment [2], but as a real possibility that may have occurred on extrasolar planets. Evolution may not produce man again, but within the scope of science we can discuss the possible convergence of some of the attributes that are characteristic of human beings. For instance, we discussed language and intelligence (cf., Chapter 6), two attributes which are of extreme importance for the search of extraterrestrial life.

What questions would a first contact with extraterrestrial life imply for both science and religion?

We have endeavored to demonstrate that contact need not come only at the level of a fully intelligent message; contact could come first in the form of detecting the first cellular steps towards intelligence; in other words, through eukaryogenesis. There remains a difficulty of addressing those attributes of human beings that are raised in theology, but not in science (spiritual dimension). The question has been formulated more precisely [16] in terms of whether a different explanation is required to understand the origins of the spiritual dimension of the human being.

While we are still not in a position to answer this question, we have endeavored to gather a number of efforts within science that suggest that contact with extraterrestrial life cannot be excluded in the future. Such an experience would give us a unique opportunity:

It would provide us with a solid point of reference on which to base original discussions of the implications on all the attributes of human beings. In such discussions the participants should be scientists and natural theologians. Facing the discussion of this possibility now is neither premature nor idle:

Exploration on Earth in the 15th century led to the difficulty of widening the horizons of the accepted attributes of man. The confrontation of Europeans with the native Americans proved to be traumatic. In retrospect, the dialogue that took place in Valladolid, Spain, between Bartolomé de las Casas, the former Bishop of Chiapas (Mexico), and the learned Juan Ginés de Sepúlveda, is still of considerable interest. The question of the attributes that characterize man was raised on that occasion.

We are still not ready to decide which attributes make us human till we reach consensus on what is our position, first on the tree of life and, subsequently, when we understand what the position of our tree of life is in the universe.

Are there trends in evolution?

In order to decide what is the position of our tree of life in the universe, we must appeal to science. First of all, we should put aside some philosophical objections that have been deeply rooted in the literature. We have touched on this question earlier, when we considered constraints on chance (cf., pp. 77-78) and evolutionary convergence (cf., pp. 161-162).

Indeed, putting together these two aspects of evolution with their intrinsic randomness suggests the following possibility:

There are trends within evolutionary history that might reflect the existence of general principles in the evolution of increasingly larger and more complex forms in the Earth biota [17] including the brain [18]. In previous chapters we have already mentioned Monod's book *Chance and Necessity*. The author overemphasized the role of 'pure chance' in evolution. He excluded the role that evolutionary convergence may have had in the evolution of life on Earth. On this basis Monod concluded that trends in biological evolution must be rejected.

This question is not merely philosophical, although its philosophical implications are important. The question of evolutionary trends is relevant to the subject of astrobiology and in particular to bioastronomy. For we have learnt in Chapter 11 that there has been an enormous technological revolution in the capability of scanning the celestial sphere for traces of ongoing communication amongst creatures that are the product of evolution of intelligent behavior elsewhere. In concluding this chapter we underline the fact that chance at the molecular level in terms of mutations in the genome, does not exclude organisms from exhibiting trends at a higher level of organization. We have already given an example of common trends at a higher level of organization in our illustrations of evolutionary convergence, which were taken form malacology and ornithology.

Another aspect of the question of the existence of trends in evolution is also relevant to natural theology, as it has been discussed extensively by Arthur Peacocke [19]. This seems to be an appropriate place to leave the subject of deeper implications of the search for extraterrestrial life at this particular cross-road of bioastronomy, philosophy and theology.

However, we would like to conclude this chapter by underlining the importance of the trends in evolution for both science and theology:

1. IN SCIENCE

The rationale for the SETI project is to assume that trends in evolution that have been observed on Earth, may serve as a basis for understanding the eventual "contact" between different forms of intelligent creatures that do not belong to the same tree of life.

2. IN NATURAL THEOLOGY.

The trends that have been observed in evolution on Earth may serve for the intrinsic and necessary problem in theology, namely, rationalizing the concept of divine action, without the fear of not being able to establish a reasonable constructive dialogue with science.

The realization that randomness in evolution does not rule out the possible existence of trends in evolution, opens the door for real progress in the integrated approach to all forms of culture. Such an approach will not fall in the trap that dates back to the publication of Darwin's seminal work, when possibly, because of the difficulty of communication between science and religion, a confrontation between faith and reason emerged.

Unfortunately, such a confrontation has not disappeared altogether.

14

Philosophical implications of the search for extraterrestrial civilizations

In the 19th century not even the advent of the theory of evolution could set speculative minds at rest regarding life's origins, either from the point of view of philosophy, or with the support of the scientific method.

Pasteur, Darwin and Wallace

The research of Louis Pasteur (1822-1895) may have had as its crowning achievement the conquest of hydrophobia. Late in his illustrious career he devoted considerable time to supervising the treatment given to patients of hydrophobia. However, during the early 1860s he had brought upon the attention of the scientific community the urgency of rationalizing the question of the origin of life. It became evident, due to Pasteur's own work, that the concept of spontaneous generation was untenable. In 1860 Pasteur wrote to a friend on the impenetrable mysteries surrounding the questions of life and death [1] ; in fact, he argued that there had been so much passion and so much obscurity on both sides of the argument that it required nothing less than the cogency of an arithmetical demonstration to convince his adversaries of the conclusions he had reached. At about the same time Charles Darwin had published *The Origin of Species*. It was only with the publication of this fundamental work that the basic questions on the nature of life would be seen in their proper perspective.

   In fact, the fundamental question of 'how life might have been breathed into matter', had to wait for two events that were to be well separated in time:

First of all, Darwin had to have his work widely read and discussed by the scientific community. The publication of his theory of evolution had taken place the previous year, although it had been maturing ever since his trip on *The Beagle*. Darwin's publication was motivated by Alfred Russell Wallace (1823-1913). His work was the fruit of travels as a naturalist on the Amazon (1848-1852) and in the Indo-Australian archipelago (1854-1862), where he independently had developed a version of evolution by natural selection.

Secondly, the first steps in the understanding of the spontaneous appearance of life from basic chemical compounds had to wait for another hundred years. It was not until the advent of chemical evolution, based on the work of Oparin, Miller and others in the middle of last century, when scientists suggested the possibility of life spontaneously appearing on Earth sometime in the remote past.

On the basis of the work of the pioneers of chemical evolution that we have mentioned in Chapter 1, philosophers of all schools were to be in a position to discuss one of the deepest questions that science has raised up to the present time, namely, What is life and how did it first appear on Earth?

Philosophical issues

We cannot repeat often enough in the present book that astrobiology's second objective is the *evolution* of life in the universe. Life on Earth is one certain example of life that can be understood in terms of evolution by means of natural selection. Hence, any account of the philosophical issues of the theory of evolution will be pertinent to astrobiology.

For this reason we should take a closer look several issues that have been discussed for a considerable time. Amongst them we would like to highlight just two, for the purpose of illustration:

• *The question of reductionism in biology.*

Reductionism, refers to any doctrine that claims to reduce the apparently more complex phenomena to the less so [2]. This will be referred to as the weak form of reductionism; its long string of successes in science, particularly in physics, is one of the success stories best known to scientists, as well as to a large sector of the educated layman. These successes can probably be best illustrated with Isaac Newton's theory of gravitation; with the introduction of quantum mechanics by Max Planck, and finally with the unified electroweak theory of Abdus Salam and Steven Weinberg (cf., Introduction).

On the other hand, there is a second meaning for the word reductionism [3]: It is the belief that human behavior can be reduced to or interpreted in terms of the lower animals, and that ultimately it can be reduced to the physical laws controlling the behavior of inanimate matter. For convenience this form of reductionism will be referred in this text as its strong version.

For example, sociobiologists and behavior geneticists have followed the tradition of the giants of physics, who restricted themselves to the weak form of reductionism). These groups of life scientists have extrapolated reductionism into its strong form. Indeed, the basis of human social behavior has been studied, in order to determine the

Indeed, the basis of human social behavior has been studied, in order to determine the relation between genetic constraints and their cultural expression. Some opponents have even referred to this approach as 'genocentrism'.

Further applications of strong reductionism have led to controversy, as clearly illustrated by Robert Russell in his recent review on *"Life in the universe: Philosophical and Theological Issues"* [4]. In particular, Russell discusses the arguments that have been put in favor of interpreting the capacity and content of human morality as products of evolution.

• *The question of design in biology.*

The question of design in biology has a long history going back to ancient Greece. However, in modern times we may begin with the work of William Paley (1743-1805), who was an archdeacon, and Doctor of Divinity at Cambridge University. His writings were highly respected in the Anglican order.

His *Horae Paulinae* (1790) was written specifically to prove the historicity of the New Testament. Another famous book was *View of the Evidences of Christianiy* (1794), a text that was standard reading amongst undergraduates during Charles Darwin early university education.

However, his best remembered book is *Natural Theology,* which played an important role in the early stages of the establishment of Darwin's arguments. Paley presented some observations from nature that were meant to prove not only the existence of a grand design, but more importantly, in his 'Natural Theology' Paley attempted to prove the existence of an intelligent designer.

The famous argument of Paley begins by supposing that he found a watch upon the ground, and someone may wonder how the watch happened to be in that place. When we come to inspect the watch, we perceive that its several parts are put together for a purpose. From this argument Paley suggested that it is inevitable that the watch must have had a maker.

This argument can be traced back to classical times, but Paley's defense of it in modern times was influential in the 19th century dialogue between science, philosophy and theology.

One of the fundamental steps in the ascent of man towards an understanding of his position in the universe (a key to understanding the present state of astrobiology), has been the realization that natural selection is indeed a creative process that can account for the appearance of genuine novelty, independent of a single act of creation, but more as a gradual accumulation of small successes in the evolution of living organisms.

This is a point that has been defended by many Darwinists, most recently by the geneticist Francisco Ayala [4]. He refers to an analogy which is artistic creation. The creative power of natural selection arises, according to Jacques Monod, from an interaction between chance and necessity, a phrase that became familiar thanks to his very popular book *"Chance and Necessity"*.

Consider a painter who mixes and distributes pigments over a canvas. The canvas and the pigments are not created by the artist, but the painting is. Leonardo's *Mona Lisa* could not be created by a random mixture of pigments, or at least the probability is infinitesimally small. This underlines the fact that natural selection is like the painter-it is not a random process. The complicated anatomy of the human eye for instance, is the result of a non-random process: natural selection.

The boundaries of science, philosophy and theology

Research in chemical evolution has provided arguments that Pasteur longed for during his lifetime. Indeed, we have attempted to be persuasive, in the sense that during the 20th century there was considerable progress in understanding the evolution of intelligent behavior on Earth.

At this stage it may be argued that our task as a scientist should be independent of that of the other approaches to the question of our origins, for [5] we may still be able to argue that philosophy is a discipline whose boundaries are in common with theology and science.

In the case of theology, discussions center on matters for which definite knowledge has, so far, been unattainable; but in the case of science discussions are based on human reason and not on assumptions beyond the realm of experiments; in other words, science avoids assumptions based on our traditions and on revelation.

On the other hand, it may be useful to be aware that this static view of the role of science has also been discussed from a different point of view; for instance, [6] it may be conjectured that science and religion are similar to each other in their search for truth.

Can there be cultural convergence?

The idea that is reflected in the last paragraph is that both science and religion are concerned with the common understanding of life in the universe. Since they largely address the same questions, both of these aspects of human culture should at some point converge.

With subsequent progress in science, philosophy and theology, convergence seems to be unavoidable. There does not seem to be any evident signs of convergence in culture at present, but the status of the relationship between the three disciplines: science, philosophy and theology (relevant to an eventual integrated view of understanding the evolution of intelligent behavior on Earth) can be discussed together, in agreement with others [7].

In fact, it cannot be denied at the beginning of the 21st century that culture demands a constant effort of synthesis of knowledge and a definite effort to integrate what we know today.

But if the extreme specialization characteristic of the basic sciences, as well as of the earth and life sciences, is not balanced by a definite effort to pay attention to the many relationships in our culture, some of which may not be immediately obvious, we are running the risk of reducing our culture to an extremely specialized activity; some may even call it "splintered culture", which goes against the original tendency of developing a common cultural background.

We may avoid a splintered culture by bringing closer to each other various approaches regarding the origin, evolution and distribution of life in the universe.

This is an area of research in which we should expect substantial progress to occur in the future. This is particularly likely, due to the large funds potentially available for space missions (cf., Table 1.1). Moreover, we are bound to converge towards a more integrated culture, due to the deep insights that all of us, scientists, philosophers and natural theologians, can collectively provide in the future.

Towards a general interest in the origin of life

The opening towards a dialogue with scientists should be seeing in the context of an ever increasing interest in questions related with the origin of life. In the United States the NASA Astrobiology Institute is a large corporation of research units that since 1998 have integrated their efforts There is a similar integration effort in Europe at the time of writing, which was evident in the first European Workshop on Exo/Astrobiology early in 2001.

Indeed, we have seen in Chapter 1 that the present time is one of expansion of the number of people that are interested in the problem of the origin of life. Theologians have been deterred from getting involved in this fascinating field, mainly due to some reservations that can be traced back a long time to the fundamental question of how to read that part of the Holy Bible which is shared by Judaism, Christianity and Islam.

In Chapter 1 we also attempted to illustrate that in the Book of Genesis there are questions raised that are of interest to theologians, philosophers, scientists and artists. To extend the fortunate phrase of Lord Snow, we may refer to these four groups as the 'four cultures'. In this terminology, it is hardly surprising that the fourth culture (art) should have shown interest in Genesis.

The very rich iconography of Christianity and Judaism was a permanent source of funding for artists throughout the rise of Western civilization. The reason why the third culture (science) has been interested in the evolution of life is self-evident. Science took one of its most transcendental steps towards understanding the complexity of the biosphere when Darwin and Wallace had formulated natural selection as a mechanism for evolution. Progress since the publication of *The Origin of Species* has been considerable. The second culture, philosophy, has been intricately connected with the development of Darwinism. Philosophers such as Karl Popper have meditated deeply on the philosophical implications of the Theory of Evolution.

We wish to dwell at a certain length on the interest that the first culture, theology, has had on the question of the origin and evolution of life on Earth and the possibility of there being life elsewhere in the universe. Traditionally, there has been a certain caution of theologians with respect to the questions that we have discussed at some length in this work. Saint Augustine touched on this question in *The City of God*[8] with respect to possible conflicts that may arise from a literal reading of the Bible.

In our own time a clear position was expressed at the Pontifical Academy of Sciences, which had met to discuss the origin and evolution of life. In this message, the subject of the present book was defined as[9] "a basic theme", which is of great interest to the Church; the idea supporting this view is that revelation contains teachings concerning the nature and origins of man.

We may raise, once again, the Augustinian rhetorical question of whether scientifically-reached conclusions, and those contained in revelation, for instance on questions related to the origin of life, are in contradiction with each other. But, at the same time, we may still ask in which direction should we seek the solution? Augustine suggests in such cases to interpret revelation allegorically. This position has opened the way to a fruitful dialogue between scientists and theologians, two cultures, which are not too distant from each other[10]. New understanding is arising between cultures that were once distant from each other. Such progress can only open the way to enhancing the new science of the origin, evolution, distribution and destiny of life in the universe (namely astrobiology) a field of research that is bound to continue its robust growth.

Is there life elsewhere in the universe?

In spite of the significant progress of the space age, the Earth is still the only place where we are certain that life has evolved, producing biodiversity which is truly remarkable: the smallest cell-walled microorganisms that have been considered by some microbiologists, the so-called 'nanobacteria', are reported to be extremely small even when compared with other microorganisms; in fact they are reported to measure only 80-500 nm[11-12]. Some of the largest living organisms, on the other hand, are the coast redwoods [for instance, *Sequoia sempervirens* (Lamb.) Endl.] that can reach up to 100 m[13]. It should be stressed that there are nine orders of magnitude between the dimensions of these two living organisms.

However, in spite of the great variety of the Earth biota and our deep knowledge of it, the question: *Is there life in space?*, still has no convincing answer. But the level, quantity and quality of research for extraterrestrial microorganisms, as well as for intelligent multicellular organisms, has increased so much in the last few years that we can only be optimistic about our eventual success.

On the persistency of biogeocentrism

It is useful to reflect on the philosophical consequences of the *biogeocentric point of view*[14], which has been advocated in the past by the evolutionist Ernst Mayr[15] and the molecular biologist Jacques Monod[16].

Mayr went as far as assigning extraterrestrial life an improbability of astronomical dimensions. His arguments were based on views related to the conditions that should be met for accepting an eventual success of the SETI project. In his opinion, each of several conditions required for the emergence of other signs of intelligent behavior are improbable. The likelihood of the presence of other civilizations would, therefore, be magnified when these improbabilities are multiplied together, according to the Drake equation (cf., Chapter 11).

On the other hand, Monod believes that the present structure of the biosphere certainly does not exclude the possibility that the decisive event occurred only once. Others have supported similar views, including the authors of "Rare Earth", which was cited in the Preface. More recently, the biogeocentric position has been maintained by the paleobiologist Simon Conway Morris[17]; his reasoning is centered on the fact that if we are alone and unique, and this is possibility which cannot be excluded at present, then this would seem to give humankind special responsibilities.

Such biogeocentric positions have also been discussed against the background of what has been called 'the principle of mediocrity'. In other words, the Copernican revolution with all its subsequent generalizations due to of Bruno, Galileo, Newton and later scientists and thinkers (cf., Chapter 15), implies that the Earth should not be regarded as a unique world.

However, independent of theoretical speculation, the question of whether we are alone in the universe will remain open until any or all of several conditions are satisfied:

Firstly, when a first evidence of intelligent behavior (i.e., communication) is detected by means of radio astronomy or by other means such as by detection of signals elsewhere in the electromagnetic spectrum (for instance in the microwave region).

Secondly, when the appropriate friendly environments for the emergence of life are directly observed (by the discovery of Earthlike extra-solar planets). This will be possible within a few years by means of telescopes in solar orbits (we have referred earlier, in Chapter 13, to the DARWIN project).

Finally, when microorganisms are detected within the our solar system. We have argued in Chapter 12 that solar system microorganisms, if they have gone beyond the simplest cellular stages, could be seen as a first step along the pathway towards the evolution of intelligent behavior.

What is the position of the human being in the cosmos?

Throughout this book we have emphasized the widespread search to understand the position of humans in the universe. In this chapter we have attempted to draw the attention of the reader to some philosophical aspects that the eventual success of such a search would imply. But to put the question of "What is the position of the human being in the cosmos?" in its proper perspective, we should carry the argument a little further:

The topic that concerns us here has also been repeatedly raised in art and literature, in particular in poetry. One striking example comes to mind, since just a few kilometers away from my desk lies the Castle of Duino, which inspired the "Duino Elegies". They began to be written during 1911-1912 by the poet Rainer Maria Rilke, and consisted of ten elegies, which were completed during the following decade [18]. They are generally concerned with man's position in the universe, as remarked earlier by Max Delbrück [19].

The question of what is our place in cosmic evolution has been raised not only across the cultural boundaries, but also the question has emerged frequently at various times in the past: almost a century and a half separate us from the concern expressed by Sir Charles Lyell regarding the position of man amongst all living organisms. As we saw in Chapter 5, it seems that the underlying difficulty barring convergence is still related to inserting Lyell's "ugly facts" of Darwinian evolution into our culture. Darwin was prudent enough to avoid ideological issues. He even avoided philosophical issues, which were inevitably raised subsequently by Bertrand Russell in *Religion and Science* [20]; for this author, from evolution no ultimately optimistic philosophy can be really inferred as far as current evidence suggests. Darwin concentrated on the narrow, but transcendental problem of establishing the theory of evolution of life on Earth; he prudently postponed the wider issue of the position of man in the universe. As we mentioned in Chapter 13, at the beginning of a new era in space research [21], a much wider problem remains to be solved, namely we still do not know what is the position of our tree of life in what may turn out to be a real *"forest of life"*.

However, we may still have to wait for preliminary progress in the still undeveloped science of the distribution of life in the universe. One of the main difficulties is deciding the position of man in our parochial tree of life. General consensus on this issue has been missing since Lyell's time.

Clearly, if we are not alone in the universe, there are some unavoidable theological and philosophical consequences for which we still have no answer (cf., Chapter 13).

However, we are convinced that discovering extraterrestrial life would induce a fruitful dialogue between sectors of culture that are not always willing to approach one another. For this reason we feel that the problem of extraterrestrial life is one of the most important questions ever raised. We hope to have conveyed to the reader the multiple hints from science that such discoveries are likely to take place sometime in the future, and possibly the near future.

The implications of such a transcendental event are likely to have an impact in our culture requiring adjustments more radical than those arising form the evidence that humans descend from microorganisms[22].

What changes can we expect in our culture if there is a prevalence of the living state throughout the cosmos?

We have concentrated in previous pages on just two subtopics of astrobiology, namely, evolution and distribution of life in the universe. What we have argued is that if the Solar System evolution experiments (cf., Chapter 8) were to be successful, the science of the distribution of life in the universe would lie on solid scientific bases. This concludes the first of the two topics I wished to comment upon.

Given such bases for the distribution of life in the universe, it does not seem premature to include in our discussions other sectors of our society. Keeping this in mind, we move on to the last topic of the present work, the destiny of life in the universe.

Indeed, if biology experiments such as the ones we have discussed in this book succeed in the future, they would undoubtedly have a significant impact in our culture, not just in our scientific outlook. The influence of the new knowledge was also felt while we were discussing the deep questions raised in Chapter 13. These questions may be discussed not just against a background of our particular evolutionary line, which has been followed up by life on Earth.

Such questions ought to be discussed already in terms of the many evolutionary lines that are hinted at by astrobiology. In fact, the intercultural dialogue is urgent for various reasons, which are strongly suggested by the main subject of this book:

• A human-level type of intelligent behavior may be widespread in the cosmos,
• It is clear that our human descent does go back all the way to microorganisms,
• The origin of microorganisms themselves goes back to star dust, which arises from supernova explosions of old stars, thus providing the stuff from which solar systems are formed, and
• Ultimately, the origin of star dust goes back to the moment in which radiation and matter separated in the aftermath of the big bang, for it was then, and only then, when atoms were finally able to form and subsequently to give rise to the condensation of star nurseries at the earliest stages of galaxy formation.

This is perhaps the most profound lesson we can infer form reading this book: cosmic evolution cannot be separated from the main concern of the new science of astrobiology: the origin evolution, distribution and eventual destiny of life in the universe. Against such a background we must persevere in the study that will take us from a proper understanding of the genesis of the living cell to the evolution of intelligent behavior in the universe.

Yet, most of these fascinating questions lie outside our cultural patrimony. But it is not an easy task to integrate our knowledge in order to avoid a "splintered culture". In this work, the main overall thesis has been that, in spite of its intrinsic difficulties, such an integrated study is not only possible, but it is also timely, necessary and urgent.

15

Back to the beginning of astrobiology

No more difficult task confronts the student of astrobiology than to trace the gradual emergence of our subject. As mentioned in Chapter 1, the basic questions that astrobiology has to face have been with us since the beginning of culture itself. At the same time the reader would have cause to complain if we had not dwelled on the origins of astrobiology, for we feel that there is no problem more interesting that this.

Astrobiology in the historical context of the sciences of the universe

The fact that we have left the origin of astrobiology till the end of the book, finds partial justification in the fact that the full range of topics that are covered spread well over the natural limits of science itself. The previous two chapters are a modest effort to integrate philosophical and theological enquiry into the scope offered to the reader.

Indeed, the destiny of life in the universe is not only one of the most fascinating areas of astrobiology, but it is also one of the least studied in depth, partly due to its interdisciplinary nature. The historical development of the key concepts in astrobiology can best be appreciated against the historical background of the whole range of sciences of the universe.

Aristarchus of Samos

The first important contribution to our present understanding of the science of astrobiology can be traced back to ancient Greece. Aristarchus of Samos lived approximately form 310 to 230 BC. He was one of the last of the Ionian scientists that founded the studies of philosophy. He lived over 23 centuries before our time, and 18 centuries before Copernicus.

In spite of such vast time gaps, Aristarchus already formulated a complete Copernican hypothesis, according to which the Earth and other planets revolve round the Sun; but in so doing, Aristarchus asserted, the Earth rotates on its axis once every 24 hours. The heliocentric theory did not prosper in antiquity. Instead, the influential astronomer Hipparchus, who flourished from 161 to 126 BC, adopted and developed a non-heliocentric theory ('epicycles'), which was going to dominate the ancient world right into the Middle Ages. The final form of the ancient model of the Solar System was defended in the middle of the second century AD by Ptolemy.

Nicholas of Cusa (Cusanus)

The Italian cardinal Cusanus (1401-1464) flourished in the middle of the 15th century AD. In 1440 he published a significant work: *De Docta Ignorantia* ("On learned ignorance"). In this book Cusanus denied the one infinite universe centered on Earth. He also claimed that the Sun was made of the same elements as Earth. He spoke of 'a universe without circumference or center'.

The work of Cusanus touched on a theological question. In his system, all celestial bodies are suns representing to the same extent the explication of God's creative power. Such dialogue between a scientific question (cosmology) and theology (divine action) would lead in the subsequent century, to a conjecture that anticipated the formulation of the central question of astrobiology: the existence of a plurality of inhabited worlds.

Nicholas Copernicus

The Polish cleric Nicholas Copernicus (1473-1543) published his heliocentric theory in 1543. During his stay at the University of Padua from 1501 to 1503, Copernicus had been influenced by the sense of dissatisfaction of the Paduan instructors with the Ptolomaic and Aristotelian systems [1].

The new century was a time in which scientists were requiring a sense of simplicity that classical philosophy could no longer provide. The main work of Copernicus, *De revolutionibus orbium coelestium* ("On the revolution of the heavenly spheres") was published a few months before his own death in 1543. In this influential work Copernicus placed the Sun at the center of the Solar System and inserted both the Sun and planets inside a sphere of fixed stars (cf., Fig. 15.1).

Copernicus deliberately appeared to accept the tenets of Aristotle's universe, as proposed in his *De Caelo,* a cosmos inserted within a system of revolving planets, surrounded by a sphere of fixed stars.

There is ample evidence that the Polish scientist knew of the heliocentric hypothesis of Aristarchus. Indeed, Bertrand Russell argues that the almost forgotten hypothesis of the Ionian philosopher did encourage Copernicus, by finding ancient authority for his innovation [2]. It would, of course, be Johannes Kepler, who was born in Germany in 1571, the scientist that would initiate a breakthrough with the discovery of the elliptical orbit. Subsequently Isaac Newton would provide the theory of gravitation that put the heliocentric theory of Copernicus on solid mathematical bases.

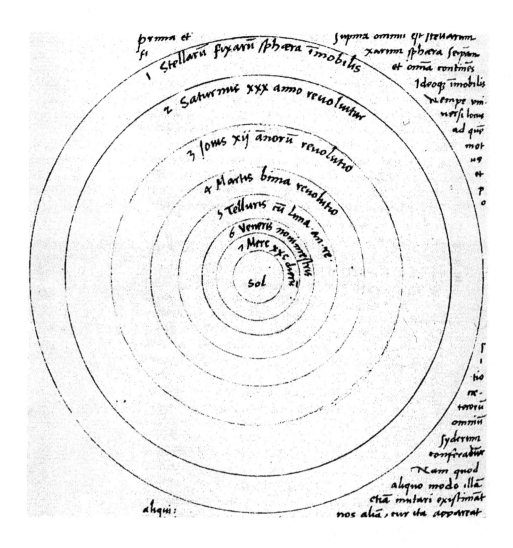

Figure 15.1: The heliocentric view of the universe according to Copernicus, in which he included a sphere of fixed stars. This last difficulty with the Copernican universe was discussed in the writings of Giordano Bruno. (Courtesy of Professor Francesco Bertola, form an original image reproduced form his book "Imago Mundi", Biblos, Cittadella PD, Italy, 1995, p. 141.)

Giordano Bruno

Giordano Bruno (1548-1600) was a philosopher, astronomer, and mathematician. He is best remembered for intuitively going beyond the heliocentric theory of Copernicus, which still maintained a finite universe with a sphere of fixed stars. As we have already seen in Chapter 10, from the point of view of astrobiology his anticipation of the multiplicity of worlds has been amply confirmed since 1995. In that year the first detection of extrasolar planets was announced. But what is more significant regarding Bruno's intuition is that he also conjectured that such worlds would be inhabited by living beings. Astrobiology, the science of life in the universe is just concerned with this key question, still without a convincing answer (cf., Figure 15.2).

Bruno was influenced by the philosophy of Cusanus (cf., previous Section). According to Hilary Gatti [3] his attention on the philosopher that preceded him was probably due to the clarity of the concepts stated by the 15th century cardinal. (An articulate expression of Bruno's reading of Cusanus can be consulted in his third Oxford Dialogue: *De l'infinito, universo e mondi ("On the infinite universe and worlds")*.

Figure 15.2: A seventeenth century representation of the Solar System, in which both planets and some of its satellites are present; but also neighboring stars are drawn with other worlds circling around them. This image represents the intuition that Bruno defended in his Italian Dialogues and in the debates at the University of Oxford. (Courtesy of Professor Francesco Bertola, form an original image reproduced   form his book "Imago Mundi", Biblos, Cittadella PD, Italy, 1995, p. 174.)

Besides, it can be said that Bruno's major innovation was his refusal to accept that the Solar System is contained in a cosmos bounded by a finite sphere of fixed stars. To sum up, Bruno proposed an infinite cosmos, populated by an infinite number of worlds. This proposal was first outlined in his first Oxford Dialogue: *The Ash Wednesday Supper*. Altogether Bruno wrote three Italian dialogues, which are relevant to our discussion. (The subsequent set of three additional dialogues refer to different matters.) His work took place during a visit to England in 1584. These writings were in fact stimulated by controversial debates at the University of Oxford. We should emphasize his third dialogue in which Bruno developed valuable concepts first introduced in *The Ash Wednesday Supper.*

His cosmological vision matured in Bruno's writings long before the science of astrometry allowed this concept to be brought within the scientific domain. Bruno's remarkable intuition is underlined by the 50 or so new extrasolar planets that were commented upon earlier, in Chapter 10. The position of Bruno was shared by Galileo Galilei in his own writings, such as the *Starry Messenger.*

Charles Darwin

As already mentioned in Chapter 13, we cannot repeat often enough that astrobiology's second objective is the evolution of life in the universe. Since life on Earth is included in its cosmic scope, any account of the history of astrobiology should take a closer look at the origin of the concept of evolution itself. Erasmus Darwin (1731-1802), a freethinker who lived through the Enlightenment, is best remembered for its two volumes under the name of *Zoonomia*, essentially a medical treatise. Charles Darwin (1809-1882) was the grandson of Erasmus.

During the Fifth Trieste Conference, the neurophysiologist Richard D. Keynes (Charles Darwin's great-grandson) presented a significant lecture in which he pointed out the possible influence of Erasmus on his grandson Charles. Indeed, Kaynes' arguments are based on the autobiography of his distinguished ancestor [4].

Indeed, Darwin had previously read the book *Zoonomia* of his grandfather, in which similar views are maintained. [Erasmus Darwin admitted the natural ascent of life and the kinship of all creatures], but without producing any effect on him. However, it is probable that the exposure rather early in life to such views, being maintained and praised in his home, may have favored his adoption of them under a different form in *The Origin of Species.*

Charles Darwin went on to complain on the degree of speculation in *Zoonomia* in proportion to the facts given. In any case the final form of Darwinism is entirely founded on the overwhelming set of supporting observations included in *The Origin of Species*. But, at the same time it is really interesting to remark that Darwin himself avoided the problem of the origin of life itself.

In his book *The Descent of Man,* which he published in 1871, Darwin maintains that it is not clear in which way mental powers were first developed in the lowest organisms. He argues that this may be as hopeless an enquiry as how life itself first originated, a question he refused to face with the same perseverance that he had demonstrated in writing of *The Origin of Species*. These were problems, according to him, for the future.

In his letters he insisted on his position with respect to the study of the origin of life. In the same year he wrote a letter to his life-long friend, the botanist Sir Joseph Dalton Hooker (1817-1911): Darwin argues that if we could imagine some 'warm little pond', supplied with chemicals, including ammonia, phosphoric salts, light, heat and electricity, then life could in principle initiate; but his argument went on to state that at the present time such primitive beginnings of life were unlikely in the presence of living organisms. This was a very reasonable attitude for the late 19th century, before experimental science began to address the question of the origin of life with Oparin and others (cf., next section). However, unconscious of the fact, Darwin influenced a young Russian scientist, who went to meet him in his home in Down House, near Farnborough in Kent, England. The name of the young scientist was K.A. Timiryazev (1843-1920). Timiryazev went on to become a Darwinist and a leading botanist, as well as Alexander Oparin's teacher. His influence on young Alexander was crucial for his pioneering work in establishing the experimental approach to the study of the origin of life on Earth [5].

Alexander Oparin

Although we have already mentioned the work of Alexander Ivanovich Oparin (1894-1980) in earlier chapters, we return to him with the question proposed by Cyril Ponnamperuma [6]: *Why was it that Oparin made such a difference?*

The answer is to be searched in the fact that Oparin was an unusual scholar, close to what sometimes is called a 'renaissance man'. He faced a problem that was basically philosophical in nature, but brought to bear upon it history as well as science. Oparin referred both to the philosophy of Aristotle, as well as to later writings of Saint Augustine. In spite of his biochemical training, Oparin was familiar with various fields that we have encountered in *The New Science of Astrobiology:* from astronomy to chemistry, from geology to biology, form philosophy to theology.

Besides the basic contributions to our field which we have referred to briefly in Chapter 1, Oparin was also a good organizer of science in his own country. In 1935 he was a co-founder (with A.N. Bakh) of the Institute of Biochemistry in Moscow, becoming its director eleven years later. Fortunately, his talent for organizing science was not restricted only to the national level. In 1957 Oparin convened the first international conference on the origin of life in Moscow. That meeting established the pattern of study of this subject. In 1963 a second international conference was held in Florida, organized by Sidney Fox (Wakulla Springs Conference).

The third international conference on the origin of life took place at Pont-á-Mousson, France. The meeting was convened by René Buvet and Cyril Ponnamperuma. On that occasion the International Society for the Study of the Origin of Life (ISSOL) was finally established. Oparin was its first President. At that time it was also decided to publish a journal on the origin of life and a newsletter. The groundwork for the foundation of ISSOL had already taken place in 1963 in Italy at the Cortina D'Ampezzo conference. Ponnamperuma recalled [6] that it was at the earlier Italian meeting when the idea first emerged of organizing the origin of life research at an international level. The role played by Oparin was essential.

Not only was Oparin a versatile scientist, the impact of his theories and experimental work laid down the foundation of the study of the origin of life as an interdisciplinary subject. His efforts opened new pathways in well established fields such as biochemistry, paleontology (especially micro-paleontology), geology, astronomy and physics.

Figure 15.3: Alexander Oparin (1894-1980, left) and Cyril Ponnamperuma (1923-1994, right) at the Cortina D'Ampezzo meeting. This event was held in 1963, after the Wakulla Springs Conference. It was organized by the International Radiation Research Conference [7]. There was a special symposium on the Origin of Life. (Photo courtesy of NASA.)

Stanley Miller

Stanley Miller was only a second year graduate student at the University of Chicago, when he published a remarkable paper on the generation of amino acids. On May 15, 1953, the journal Science published a seminal paper, which consisted of a brief note, whose title was "A Production of Amino Acids under Possible Primitive Earth Conditions".

This publication reported a simple experiment, which attempted to reproduce conditions similar to those in the early Earth, when life first originated. Miller was under the guidance of Harold Urey (cf., Chapter 4, "The Primitive Earth"), who had done fundamental research in nuclear physics. Urey was responsible for the discovery of an isotope of hydrogen: deuterium. He received the Nobel Prize for this work.

Urey had subsequently suggested that the early Earth had conditions favorable for the formation of organic compounds. Urey had become interested in the early Earth atmosphere. He conjectured that it was formed out of a mixture of several gases, prominent amongst which were molecular hydrogen, methane, ammonia and water vapor. As a subject of his doctoral thesis Miller demonstrated experimentally that amino acids, the building blocks of the proteins, could be formed without the intervention of man in environmental conditions, which in earlier chapters we have called prebiotic - gases similar to those that presumably were present at the earliest stages in the evolution of the Earth itself.

Miller investigated the effect of lightning on the proposed primitive atmosphere. To everyone's delight, once the mixture received repeated electric discharges, the water within the closed glass recipient that contained the Urey candidate gases for the early atmosphere, was condensed. (It was no longer pure water.) Chemical analysis demonstrated that amino acids had been synthesized. The corresponding geologic period in which chemical synthesis of biomolecules occurred was referred to in chapters 2 - 4 as the Archean.

Miller's work was a significant step in the growth of the subject of chemical evolution.

Recent studies of the early Earth atmosphere suggest that the more abundant primitive gases may have been carbon dioxide and molecular nitrogen, but nevertheless estimates of the primitive gases are still uncertain, and the Miller experiment remains a landmark in the development of astrobiology.

Throughout his long and productive career Miller has continued to inspire younger generations with his research on a wide spectrum of topics on chemical evolution. His presence at the Trieste Conference in September, 2000, where he delivered the Evening Lecture on: "Peptide nucleic acids as a possible primordial genetic polymer", was undoubtedly an inspiration to those present for the occasion.

Sidney W. Fox

Sidney Fox (1912 - 1998) brought vitality and excitement to the work on the origin of life. He participated in three of the Trieste Conferences (1994, 1995 and 1997). Fox was particularly well appreciated for many reasons:

In spite of the presence amongst many researchers at the conference site in Miramare, Fox stood out as a uniquely persistent scientist. He had been making significant contributions valuable to the field of the origin, evolution and distribution of life in the universe for over half a century. The publication of his sequence analysis of amino acid residues in 1945 was to be followed by an enormous output of influential papers, which were written in collaboration with over sixty associates.

The study of the origin of life has gained many insights due to his work. He was the first to synthesize a protein by heating amino acids under conditions found here on Earth. He also showed that these new thermal proteins, when placed in water, would self-organize into a primitive cell.

Figure 15.4: Sidney Fox (1912-1998) during his first visit to the Fourth Trieste conference in 1995. (Courtesy of ICTP Photo Archives.)

Fox was the first vice president of ISSOL. He had urged Oparin, its first president, that the time was ripe for a society that would bring together specialists from many countries and disciplines. He organized numerous conferences including, as we have mentioned above, the Wakulla Springs Conference in 1963. This meeting brought together both pioneers of the studies of the origin of life: Oparin and the English biologist John Burdon Sanderson Haldane (1892-1964).

Sidney Fox also produced many books related to the origin of life. By persevering in the field of chemical evolution for so long, he will be remembered as one of the chief pioneers of the emerging field of astrobiology.

Cyril Ponnamperuma

Cyril Ponnamperuma (1923-1994) was born in Ceylon, now Sri Lanka, and went to the USA in 1959. Before his arrival, while studying at the University of London, he had come under the influence of the physicist John D. Bernal (1901-1971), a well-known crystallographer who had published in 1949 an early, but influential paper on the idea that the organic compounds that serve as a the basis of life, were formed when the Earth had an atmosphere of methane, ammonia, water and hydrogen.

Ponnamperuma came in contact with the Nobel Laureate Melvin Calvin (1911-1997) in 1962. This led to a series of papers on the synthesis of DNA components, extending in a significant way the pioneering work initiated by Stanley Miller.

In 1971, he joined the Maryland faculty as head of the Laboratory of Chemical Evolution, which he directed until his death. He became principal organic analysis investigator for the Apollo project and also worked on the Viking and Voyager programs. He left NASA to join the Maryland faculty. Ponnamperuma played a major role in NASA's early experiments for detecting life on Mars at the time of the Viking missions (cf., Chapter 7). His analysis of meteorites showed that the basic chemicals of life were not confined to the Earth. In the analysis of the meteorite that fell in Murchison, Australia in 1969 (cf., Chapter 3), he and his co-workers provided evidence for extraterrestrial amino acids and hydrocarbons.

He played a leading role in ISSOL from 1977 till 1986, He was the recipient of awards from several countries. These included the first Alexander Ivanovich Oparin Medal awarded by ISSOL at its triennial meetings for the best "sustained scientific research program" in the origin of life field (in 1980). Since 1990 he was closely associated with the ICTP. He interacted with Abdus Salam in research on the chirality of amino acids, and tested some of these new ideas in his own Laboratory of Chemical Evolution. This collaboration was the initiation of a longer association with the ICTP in the form of the series of conferences on various aspects of astrobiology, which we have already mentioned in the Preface .

The recent period

At present the new science of astrobiology is developing at a fast pace. It is virtually impossible in a general book like the present one, whose main purpose is not historical, to go beyond this brief sketch of the influence of a few researchers that contributed significantly to lay down the bases of astrobiology.

It has been a great privilege for our generation to share with some of these pioneers our interest in life: its origin, evolution, distribution and destiny. The present book is to a large extent the result of what the author has learnt at the Trieste and Caracas meetings, where all the participants who attended the events in both cities had the good fortune to interact with the above scientists and many other pioneers of the new science of astrobiology.

16

Recapitulation

One of our objectives has been to present the new science of astrobiology in a social and cultural context.

The Central Questions of Astrobiology

The initial work for this book began in 1996. It was intended as a set of notes for a lecture to which I was kindly invited by the International School of Plasma Physics "Piero Caldirola" (Varenna, Lake Como, Italy). On that occasion the School chose the topic: "Reflections on the Birth of the Universe: Science, Philosophy and Theology" [1].

The 43 pages of the Varenna lecture were subsequently considerably modified thanks to the kind invitation of the directors of the third volume of the series on *Scientific Perspectives on Divine Action,* which dealt with the subject of "Evolutionary and Molecular Biology: Scientific Perspectives on Divine Action" [2]. That activity was co-sponsored by the Vatican Observatory, Vatican City State, and the Center for Theology and the Natural Sciences in Berkeley, California. The book was published in 1998. In the following three years I have endeavored to extend the scope of my presentation of the new science of astrobiology.

I have attempted, as far as possible, to organize the multiple topics that should be covered in a text facing the almost impossible task of sketching the main aspects of astrobiology, namely:

- the origin,
- evolution,
- distribution and
- destiny of life in the universe.

The result of my efforts has been presented in the previous 15 chapters:

185

In BOOK 1 we began the first of the major topics of astrobiology, the origin of life in the universe. We introduced chemical evolution as a process that comes at the end of cosmic evolution (Chapter 1). This, in turn, takes us to the study of prebiotic evolution (Chapter 2). We also undertook a brief evaluation of the sources of the possible precursors of the macromolecules that would enter a chemical evolution pathway towards the macromolecules of life.

In BOOK 2 we introduced the topic of evolution of life in the universe, by discussing the processes that have taken life on Earth from the age of the prokaryotes to the origin of eukaryotes and, eventually, to the evolution of intelligent behavior. This work was covered in preparation for the introduction of the problem of searching for the evolution of intelligent behavior on other worlds, a topic which was developed in chapters 10 to 12.

In BOOK 3 we took up the distribution and destiny of life in the universe. We have restricted ourselves to the main successes of exobiology: the study of Mars, Europa and Titan. In the second part of Book 3, we touched upon bioastronomy, which deals with the study of environments for the possible distribution of beings that might be the product of evolution of intelligent behavior on other worlds.

The concluding section of the book included three general topics: firstly, the search for new planets in other solar systems; secondly, a brief review of the programs in radio astronomy that are searching for extraterrestrial life and, finally, the exploration of what we know about life on Earth, in order to answer one of the central questions of astrobiology, namely: *Is the evolution of intelligent behavior universal?*

The final chapters were reserved for the deeper cultural implications of astrobiology, concerning the destiny of life in the universe. We divided the topics discussed into theological, philosophical and historical issues that are relevant to astrobiology.

BOOK 4:

SUPPLEMENT

Notes and references

PREFACE

1. Bertola, F. (1992) Seven centuries of astronomy in Padua, in *From Galileo to the Stars,* Biblos Edizione, Cittadella (PD), Italy.

2. Chela-Flores, J. (1997) Cosmological models and appearance of intelligent life on Earth: The phenomenon of the eukaryotic cell, in Padre Eligio, G. Giorello, G. Rigamonti and E. Sindoni (eds.), *Reflections on the birth of the Universe: Science, Philosophy and Theology,* Edizioni New Press, Como, pp. 337-373.

3. Chela-Flores, J. (1998) The phenomenon of the eukaryotic cell, in R. J. Russell, W. R. Stoeger and F. J. Ayala (eds.), *Evolutionary and Molecular Biology: Scientific Perspectives on Divine Action.* Vatican City State/Berkeley, California, Vatican Observatory and the Center for Theology and the Natural Sciences, pp. 79-99.

4. Chela-Flores, J. (1999a) Search for the Ascent of Microbial Life towards Intelligence in the Outer Solar System, in Roberto Colombo, Giulio Giorello and Elio Sindone (eds.), Edizioni New Press, Como.

5. Chela-Flores, J. (2000) *Testing the Drake Equation in the Solar System, A New Era in Astronomy,* Lemarchand G.A. and Meech K. (eds.), ASP Conference Series, San Francisco, USA, **213,** 402-410.

6. Chela-Flores, J. (1999b) Gli alberi della vita, in Elio Sindone and Corrado Sinigaglia (eds.), *Carlo Maria Martini, Orizzonti e limiti della scienza,* Decima Cattedra dei non credenti, Milano, Raffaello Cortina Editore, Milano, pp. 43-50.

INTRODUCTION

1. We refer the reader to the Glossary for the precise definition of "taxonomy"; cf., also Mayr, E. (1982) *The Growth of Biological Thought. Diversity, Evolution and Inheritance,* Belknap Press/Harvard University Press, Cambridge, Mass.

2. Darwin, C. (1859) *The origin of species by means of natural selection or the preservation of favored races in the struggle for life.* London, John Murray. Reprinted by Oxford World's Classics (1998), Gillian Beer (ed.), Oxford University Press.

3. Fischer, E. P. and Lipson, C. (1988) *Thinking about Science (The life of Max Delbruck).* W.W. Norton, New York.

4. Schrodinger, E. (1967) *What is life?,* Cambridge University Press.

5. Wolpert, L. (1990) *New Scientist* **125** No. 1707, 64.

6. Cf. Glossary: "cyanobacteria", "stromatolites", "photosynthesis"; cf., also Schopf, J.W. (1993), Microfossils of the Early Archean Apex Chert: New Evidence of the Antiquity of Life. Science **260,** 640-646.

7. Cf. Glossary: "geochronology"; cf., also Schidlowski, M. (1995) Early Terrestrial Life: Problems of the oldest record, in Chela-Flores, J., M. Chadha, A. Negron-Mendoza, and T. Oshima (eds.). (1995) *Chemical Evolution: Self-Organization of the Macromolecules of Life,* A. Deepak Publishing, Hampton, Virginia, USA, pp. 65-80;. Mojzsis, S. J., Arrhenius, G., McKeegan, K.D., Harrison, T.M., Nutman, A.P. and Friend, C.R. (1996) Evidence for life on Earth before 3,800 million years ago, *Nature* **384,** 55-59; Moorbath, S. (1995) *Age of the oldest rocks with biogenic components,* in Ponnamperuma, C. and Chela-Flores, J. (eds.). (1995) *Chemical Evolution: The Structure and Model of the First Cell,* Kluwer Academic Publishers, Dordrecht, The Netherlands. pp. 85-94.

8. Cf. Glossary: "Archean", "isotope".

9. Margulis, L. and Sagan D. (1987) *Microcosmos Four Billion Years of Evolution form Our Microbial Ancestors,* Allen & Unwin, London, pp. 99-114.

10. Salam, A. (1991) The role of chirality in the origin of life, *J. Mol. Evol.,* **33,** 105-113 and Chirality, phase transitions, and their induction in amino acids, *Phys. Lett.* **B288,** 153-160; Chela-Flores, J. (1991) Comments on a Novel Approach to the Role of Chirality in the Origin of Life, *Chirality* **3,** 389-392; Chela-Flores, J. (1992) The origin of chirality in protein amino acids, *Chirality* **6,** 165-168.

11. This remark is true for eubacterial cell walls, but strictly speaking it cannot be considered an exception to protein synthesis, because these peptides with D-amino acids are not synthesized at the level of the ribosomes. cf. also Glossary, "amino acids", "phospholipids", "biochemistry", "chirality" and the "genetic code".
12. Cf. Glossary: under "weak interaction".

13. Cf. Glossary: under "parity violation".

14. Cf. Glossary: under "quantum chemistry".

15. Ponnamperuma, C. and Chela-Flores, J. (eds.), (1993) *Chemical Evolution: Origin of Life,* A. Deepak Publishing, Hampton, Virginia, USA.

16. Chela-Flores, J., Chadha, M., Negron-Mendoza, A. and Oshima, T. (eds.), (1995) *Chemical Evolution: Self-Organization of the Macromolecules of Life,* A. Deepak Publishing, Hampton, Virginia, USA.

17. Ponnamperuma, C. and Chela-Flores, J. (eds.), (1995) *Chemical Evolution: The Structure and Model of the First Cell,* Kluwer Academic Publishers, Dordrecht.

18. Chela-Flores, J. and Raulin, F. (eds.), (1996) *Chemical Evolution: Physics of the Origin and Evolution of Life,* Kluwer Academic Publishers, Dordrecht.

19. Chela-Flores, J. and Raulin, F. (eds.), (1998) *Chemical Evolution: Exobiology. Matter, Energy, and Information in the Origin and Evolution of Life in the Universe,* Kluwer Academic Publishers, Dordrecht.

20. Chela-Flores, J., Owen, Tobias and Raulin, F. (eds.), (2001) *The First Steps of Life in the Universe.* Kluwer Academic Publishers: Dordrecht, The Netherlands. (In preparation.)

21. Chela-Flores, J., Lemarchand, G. A. and Oro, J. (eds.), (2000) *Astrobiology From the Big Bang to Civilization,* Kluwer Academic Publishers, Dordrecht.

CHAPTER 1.

FROM COSMIC TO CHEMICAL EVOLUTION

1. Russell, B. (1991) *History of Western Philosophy and its Connection with Political and Social Circumstances from the Earliest Times to the Present Day,* Routledge, London, p.13.

2. Coyne, S. J., G. (1996) *Cosmology: The Universe in Evolution,* in Chela-Flores and Raulin (1996), loc. cit., pp. 35-49.

3. Kennicutt Jr., R. C. (1996) *Nature* **381,** 555-556. Subsequent work with the HST (Wendy Freedman and co-workers) tends to confirm this value of H_O .

4. This section is partially based on three reviews; firstly, Caldwell, R.R. and Kamionkowski (2001). *Echoes from the Big Bang,* Scientific American January 2001, pp. 28-33, secondly, Rago, H. (2000).*Cosmos and Cosmology,* in Chela-Flores, J., Lemarchand, G. A. and Oro, J. (eds.), (2000) *Astrobiology From the Big Bang to Civilization,* Kluwer Academic Publishers, Dordrecht, pp. 33-40; finally, Sneden C. (2001) *The age of the universe* Nature **409,** 673-674.

5. Cf. Glossary: "supernova"; cf., also Riess, A., *et al* (1998) Observational evidence from supernovae for an accelerating universe and a cosmological constant, *Astronomical Journal* **116,** 1009-1038.

6. Krauss, L. M. (1998) *The end of the age problem and the case for a cosmological constant revisited.* Astrophysical Journal 501, 461-466; Ostriker, J.P. and Steinhardt, P.J. (2001). *The Quintessential Universe,* Scientific American January 2001, pp. 37-43.

7. Cf. Shapiro, R. (1996) *Prebiotic synthesis of the RNA bases. A critical analysis,* Summary presented at the ISSOL Conference, Orléans, July, Book of Abstracts, p. 39.

8. Ponnamperuma, C. (1995) *The origin of the cell from Oparin to the present day,* in Ponnamperuma, C. and Chela-Flores, J. (eds.), (1995) *Chemical Evolution: The Structure and Model of the First Cell,* Kluwer Academic Publishers, Dordrecht, pp. 3-9.

9. Darwin, C. (1859) *The origin of species by means of natural selection or the preservation of favored races in the struggle for life,* John Murray/Penguin Books, London, 1968, p. 455.

10. Woese, C. R. (1983) The primary lines of descent and the universal ancestor, in Bendall D.S. (ed.), *Evolution from Molecules to Man,* Cambridge University Press, London, pp. 209-233; cf., also Islas, S., Becerra, A., Leguna, J. I. and Lazcano, A. (1998) Early cellular evolution: Preliminary results from comparative genomic analysis, in Chela-Flores, J. and Raulin, F. (eds.), (1998) *Chemical Evolution: Exobiology: Matter, Energy, and Information in the Origin and Evolution of Life in the Universe,* Kluwer Academic Publishers, Dordrecht, pp.167-174.

11. Cf. Glossary: "DNA, "RNA".

12. Coveney, P. and Highfield, R. (1995) *Frontiers in Complexity. The search for order in a chaotic world,* Faber and Faber, London, pp. 190-236.

13. Meyer, M. A. (1996) The search for life and its origins: Why NASA?, Poster 3, 8th ISSOL Meeting. 11th International Conference on the Origin of Life, Orléans, France, July 8-13, Book of Program and Abstracts, p. 63.

14. Since the late 1990s comparative planetology has been enriched by many extrasolar planets as well as by new solar systems, such as the one discovered in Upsilum Andromedae, a star which belongs to the Andromeda constellation (cf., Chapter 10). For an earlier view on comparative planetology the reader should consult Murray, B., Malin, M.C., and Greely, R. (1981) *Earthlike Planets. Surfaces of Mercury, Venus, Earth, Moon, Mars,* W. H. Freeman & Co., San Francisco, p. 317.

15. Drake, F. and Sobel, D. (1992) *Is there anyone out there? The scientific search for Extraterrestrial Intelligence,* Delacorte Press, New York.

16. Heidmann, J. (1996) *SETI from the Moon. A case for a 21st Century SETI-Dedicated Lunar Farside Crater,* in Chela-Flores, J. and Raulin, F. (eds.), *Chemical Evolution: Physics of the Origin and Evolution of Life,* Kluwer Academic Publishers, Dordrecht, pp. 343-353. The work of the late Heidmann has been carried on by Giancarlo Genta (cf., Proceedings of the Sixth Trieste Conference on Chemcal Evolution. (In preparation.)

17. Cf. Glossary: "chemosynthesis", "supernova".

18. Ponnamperuma, C. (1993) The origin, evolution and distribution of life in the universe, *in* Ponnamperuma, C. and Chela-Flores, J. (eds.), (1993) *Chemical Evolution: Origin of Life,* A. Deepak Publishing, Hampton, Virginia, USA, p. 6.

19. Cf. Glossary: "Orion Nebula".

20. Cf. Glossary "angular momentum".

21. Pollack, J. B. and Atreya, S. K. (1992) *Giant planets: Clues on Current and Past Organic Chemistry in the Outer Solar System,* in G.C. Carle, D.E. Schwartz, and J.L. Huntington, (eds.), *Exobiology in Solar System Exploration,* NASA publication SP 512. p.96;

22. Cf. Glossary: "hydrocarbons".

23. cf., Abbreviations: AU, HIPPARCOS, HST, Myr; cf., also Glossary, "Kuiper Belt", "zodiacal dust cloud"; finally we refer the reader to Schneider, G., Smith, B. A., Becklin, E. E., Koerner, D. W., Meier, R. Hines, D. C., Lowrance, P. J., Terrile, R. J., Thomson, R. I. and Rieke, M. (1999) NICMOS Imaging of the HR 4796A Circumstellar Disk, *Astrophys. Journal* **513,** L127-L130.

24. This parameter - the surface temperature - is indicated by what is technically known as its 'Planck radiation', the name given in honor of the Physics Nobel Laureate Max Planck, to the distribution with wavelength of the radiation emitted by the star at various temperatures. Such temperatures can be measured by the spectral type of the star, or by its color index. (cf., Glossary for definitions of spectral type and color index.)

25. Cf. Glossary: "photosphere".

26. Cf. Chapter 10, section: "Which are likely habitable zones?".

27. Cf. Chapter 5, "evolution of the hominoids".

28. Wells, H. G. (1996) *The Time Machine,* Phoenix, London, Abridged edition.

CHAPTER 2.

FROM CHEMICAL TO PREBIOTIC EVOLUTION

1. Cf. Glossary: "amino acids", "bases".

2. Cf. Glossary: "interstellar dust".

3. Cf. Glossary: "photosphere"; cf., also Fegley Jr., B. (1993) Chemistry of the Solar Nebula, in *The Chemistry of Life's Origins,* J.M. Greenberg, C.X. Mendoza-Gomez

and Piranello, V. (eds.), Kluwer Academic Publishers, Dordrecht, pp. 75-147; N. Grevesse (1984) Accurate atomic data and solar photospheric spectroscopy, *Physica Scripta* **T8**, 49-58.

4. To put meteorite composition on a convenient scale, we have adopted the standard normalization, in which an abundant element in the silicates (a 'lithophile') with high melting temperature is chosen; in the present case it is Si. Other possibilities are Mg or Al. In the solar photosphere the standard choice is the element hydrogen (cf., Lipschutz, M. E. and Schultz, L. (1999) Meteorites, in *Encyclopedia of the Solar System,* P. R. Weissman, L.-A. McFadden and T. V. Johnson (eds.), Academic Press, San Diego. pp. 629-671. For meteorites the data are presented on a weight basis, such as in the Elsevier Table, where the results are given in ppm by weight; cf., Lof, P. (ed.), (1987) *Elsevier's Periodic Table of Elements*, Elsevier Science Publishers, Amsterdam. Alternatively the results are given on atom basis, as we have done in Table 2.1. For the hydrogen abundance in the CI chondrite, we have used the specific value for the Orgueil chondrite (where we have equated the carbon abundance of 34,500 ppm by weight with the atomic scale value cited in Table 2.1).

5. Oro, J. (1995) Chemical synthesis of lipids and the origin of life, in Ponnamperuma, C. and Chela-Flores, J. (eds.), (1995) *Chemical Evolution: The Structure and Model of the First Cell,* Kluwer Academic Publishers, Dordrecht, pp. 135-147.

6. Oro, J. (1961) Comets and the formation of biochemical compounds on the primitive earth, *Nature* **190,** 389-390.

7. Brownlee, D. E. and Sandford, S. A. (1992) Cosmic Dust, in G. C. Carle, D. E. Schwartz, and J. L. Huntington, (eds.), *Exobiology in Solar System Exploration,* NASA Publication SP 512, pp. 145-157.

8. Cf. Glossary: "unsaturated", "hydrocarbons"; cf. also Delsemme, A. H. (1992) *Comets: Role and importance to Exobiology,* in G. C. Carle, D. E. Schwartz, and J. L. Huntington (eds.), *Exobiology in Solar System Exploration.* NASA publication SP 512, pp. 177-197.

9. Cf. Chapter 1, where "The Origin of Elements" was discussed.

10. For an up-to-date review, see Cassen, P. and Woolum, D. S. (1999) *The origin of the Solar System,* in P. R. Weissman, L.-A. McFadden and T. V. Johnson (eds.), *Encyclopedia of the Solar System,* Academic Press, San Diego, pp. 35-63.

11. Cf. Table 8.4.

12. Cf. Glossary: "terrestrial planets.

13. Cf. Glossary: "carbonaceous chondrites".

14. Cf. Glossary: "silicate".

15. The reader should keep in mind that the discoveries of planets outside our solar system may require deeper insights than those sketched in Chapter 2; cf. Glossary:

"Jovian planets". In particular, the Jovian density is 1.3 gm/cm^3. This should be compared with the terrestrial density which is 5.5 gm/cm^3.

16. Cf. Glossary: "accretion".

17. Pollack, J. B. and Atreya, S. K. (1992) *Giant planets: Clues on Current and Past Organic Chemistry in the Outer Solar System,* in G. C. Carle, D. E. Schwartz, and J. L. Huntington (eds.), *Exobiology in Solar System Exploration,* NASA publication SP 512, pp. 83-101; cf. also ref. 1, Chapter 3.

CHAPTER 3.

SOURCES FOR LIFE'S ORIGINS: A SEARCH FOR BIOGENIC ELEMENTS

1. For the definition of albedo we refer the reader to the Glossary, whereas for further physical parameters the appropriate reference is: Ross Taylor, S. (1999) *The Moon,* in P. R. Weissman, L.-A. McFadden and T. V. Johnson (eds.), *Encyclopedia of the Solar System,* Academic Press, San Diego, pp. 247-275.

2. Ponnamperuma C. (1995) The origin of the cell from Oparin to the present day, in Ponnamperuma, C. and Chela-Flores, J. (eds.), (1995) *Chemical Evolution: The Structure and Model of the First Cell,* Kluwer Academic Publishers, Dordrecht, pp. 3-9.

3. Cf. Glossary: "breccia, "basalt and "arnothosite"; a remarkable Apollo 16 sample of arnothosite was assigned an age of approximately 4440 Myr; cf., also Gibson,, E. K. and Chang, S. (1992) The Moon: Biogenic elements, in G. C. Carle, D. E. Schwartz and J. L. Huntington (eds.), *Exobiology in Solar System Exploration,* NASA SP 512, pp. 29-43.

4. Britt, D. T. and Lebofsky, L. A. (1999) Asteroids, in P. R. Weissman, L.-A. McFadden and T. V. Johnson (eds.), *Encyclopedia of the Solar System,* Academic Press, San Diego, pp. 585-605.

5. Campins, H. (2000) The chemical composition of comets, in Chela-Flores, J., Lemarchand, G.A. and Oro (eds.), J. (2000). *Astrobiology: Origins from the Big Bang to Civilisation,* Kluwer Academic Publishers, Dordrecht, The Netherlands, pp. 163-176.

6. McSween, H. Y and Stolper, E. M. (1980) *Basaltic meteorites, Scientific American* **242**, Number 6, pp. 44-53.

7. Monod, J. (1972) *Chance and Necessity An Essay on the Natural Philosophy of Modern Biology,* Collins, London.

8. Oro, J. Squyres, S. W., Reynolds, R. T., and Mills, T. M. (1992) Europa: Prospects for an ocean and exobiological implications, in G. C. Carle, D. E. Schwartz and J. L. Huntington (eds.), *Exobiology in Solar System Exploration,* NASA SP 512, pp. 103-125.

9. Orgel, L. (1994) The origin of life on Earth, *Scientific American* **271**, No. 4, 53-61.

10. Wolman, Y., Haverlard, W. J., and Miller, S. L. (1972) Non-protein amino acids from spark discharges and their comparison with Murchison meteorite amino acids, *Proc. Natl. Acad. Sci. USA* **69**, 809-811.

11. Kasting, J. F. (1993) Earth's Early Atmosphere, *Science* **259**, 920-926.

12. For an explanation of the term 'biomarker' we refer the reader to the Glossary. The original discussion of the Allan Hills meteorite is due to McKay, D. S., Gibson Jr., E. K., Thomas-Keprta, K. L., Vali, H., Romanek, C.S., Clemett, S. J., Chiller, X. D. F., Maechling, C. R., and Zare, R. N. (1996), Search for past life on Mars: Possible Relic Biogenic Activity in Martian Meteorite ALH84001, *Science* **273**, 924-930.

13. Cf. Glossary: "steranes" and "steroids"; cf. also Chadha, M. (1996) Role of transient and stable molecules in chemical evolution, in Chela-Flores, J. and Raulin, F. (eds.), *Chemical Evolution: Physics of the Origin and Evolution of Life,* pp. 107-122.

14. Doran, P. T., Wharton, Jr., R. A. and Berry Lyons, W. (1994) Paleolimnology of the McMurdo Dry Valleys, Antarctica, *J. Paleolimnology* **10**, 85-114.

15. Wharton, Jr., R. A., Parker, B. C. and Simmons, Jr., G. M. (1983) Distribution, species composition and morphology of algal mats in Antarctic dry valley lakes, *Phycologia* **22**, 355-365.

16. Parker, B. C., Simmons, Jr., G. M., Wharton, Jr., R. A., Seaburg, K. G. and Gordon Love, F. (1982) Removal of organic and inorganic matter from Antarctic lakes by aerial escape of bluegreen algal mats, *J. Phycol.* **18**, 72-78.

17. Parker, B. C., Simmons, Jr., G. M., Gordon Love, F., Wharton, Jr., R. A. and Seaburg, K. G. (1981) Modern Stromatolites in Antarctic Dry Valley Lakes, *BioScience* **31**, 656-661.

18. Ellis-Evans, J. C. and Wynn-Williams, D. (1996) A great lake under the ice, *Nature* **381**, 644-646.

19. Priscu, J.C., Adams, E.E., Lyons, W.B., Voytek, M.A., Mogk, D.W., Brown, R.L.. McKay, C.P., Takacs, C.D., Welch, K.A., Wolf, C.F., Krishtein, J.D., and Avci, R. (1999) Geomicrobiology of subglacial ice above Lake Vostok, Antartica, *Science* **286**, 2141-2144.

CHAPTER 4.

FROM PREBIOTIC EVOLUTION TO SINGLE CELLS

1. More precisely, the angle a through which the plane of polarization is rotated can be given a sign as follows: when looking towards the incoming beam of light, for a clockwise rotation of a, we would assign the term 'right-handed'; if such a rotation is anticlockwise we say that the rotation is 'left-handed'.

2. Oró, J. (1995) *Chemical synthesis of lipids and the origin of life,* in Ponnamperuma, C. and Chela-Flores, J. (eds.), (1995) *Chemical Evolution: The Structure and Model of the First Cell,* Kluwer Academic Publishers, Dordrecht, pp. 135-147.

3. Cf. Glossary: "glycerol"; cf. also Kandler, O. (1995) Cell wall biochemistry in Archaea and its phylogenetic implications, in Ponnamperuma, C. and Chela-Flores, J. (eds.), (1995) *Chemical Evolution: The structure and model of the first cell,* Kluwer Academic Publishers, Dordrecht, pp. 165-169.

4. Cf. Glossary: "neutron star, "supernova and enantiomer; cf. also Chapter 2, "molecular clouds".

5. Bonner, W. A. (1991) The origin and amplification of biomolecular chirality, *Origins of Life and the Evolution of the Biosphere* **21**, 59-111.

6. Greenberg, J. M., Kouchi, A., Niessen, W., Irth, H., van Pardijs, J., de Groot, M., and Hermsen, W. (1995) *Interstellar dust, chirality, comets and the origins of life: Life from dead stars? ,* in Ponnamperuma, C. and Chela-Flores, J. (eds.), (1995) *Chemical Evolution: The Structure and Model of the First Cell,* Kluwer Academic Publishers, Dordrecht, pp. 61-70.

7. Cline, D., Liu, Y. and Wang, H. (1995) Effect of a chiral impulse on the weak interaction induced handedness in a prebiotic medium, *Origins of Life and the Evolution of the Biosphere* **25**, 201-209.

8. The Murchison meteorite will be discussed more fully in Chapter 6; cf. also Oro, J., Miller, S. L., and Lazcano, A. (1990) The origin and early evolution of life on Earth *Ann. Rev. Earth Planet Sci.* **18**, 317-356.

9. Ponnamperuma C. (1995) *The origin of the cell from Oparin to the present day,* in Ponnamperuma, C. and Chela-Flores, J. (eds.), (1995), *Chemical Evolution: The Structure and Model of the First Cell,* Kluwer Academic Publishers, Dordrecht, pp. 3-9.

10. Cronin, J. R. and Pizzarello, S. (1997) *Enantiomeric excesses in meteoritic amino acids, Science* **275**, 951-955.

11. Bonitz, S. G., Berlani, R., Corruzzi, G., Li, M., Macino, G., Nobrega, F. G., Nobrega, M. P., Thalenfeld, B. E., and Tzagoloff, A. (1980) Codon recognition rules in yeast mitochondria, *Proc. Natl. Acad. Sci. USA* **77**, 3167-3170.

12. Bruijn, M. H. L. (1983) *Drosophila melanogaser* mitochondrial DNA, a novel organization and genetic code, *Nature* **304**, 234-241.

13. Barrel, B. G., Bankier, A.T. and Drouin, J. (1979) A different genetic code in human mitochondria, *Nature* **282**, 189-194.

14. Margulis, L. (1993) *Symbiosis in Cell Evolution,* Freeman & Co., San Francisco.

15. Margulis, L. and Sagan, D. (1987) *Microcosm,* Allen & Unwin, London, p. 132.

16. We have attempted to be consistent in our reference to stratigraphy by referring to the conventions of J. W. Schopf (ed.), (1983) *Earth's Earliest Biosphere Its Origin and Evolution.* Princeton University Press, Princeton.

17. Sleep, N. H., Zahnle, K.J., Kasting, J. F., and Morowitz, H. J. (1989) Annihilation of ecosystems by large asteroid impacts on the early Earth *Nature* **342**, 139-142.

18. Schidlowski, M. (1995) Early terrestrial life: Problems of the oldest record, in Chela-Flores, J., Chadha, M. Negron-Mendoza, A. and Oshima, T. (eds.), *Chemical Evolution: Self-Organization of the Macromolecules of Life,* A. Deepak Publishing: Hampton, Virginia, USA. pp. 65-80

19. Schopf, J. W. (1993) Microfossils of the Early Archean Apex Chert: New Evidence of the Antiquity of Life, *Science* **260**, 640-646.

20. Mojzsis, S.J., Harrison, T.M. and Pidgeon, R.T. (2001) Oxygen-isotope evidence from ancient zircons for liquid water at the Earth's surface 4,3000 Myr ago, *Nature* **409**, 178-181; cf., also Halliday, A.N. (2001) *In the beginning...*, *Nature* **409**, 144-145.

21. Chatton, E. 1937; for a review of the prokaryotic and eukaryotic dichotomy cf. Rizzotti, M. (2000) *Early Evolution,* Birkhauser Verlag, Basel, Chapter 3, pp. 24-52.

22. Zukerkandl, E. and Pauling, L. (1965) Molecules as documents of evolutionary history, *J. Theor. Biol.* **8**, 357-366.

23. Woese, C. R. (1983) The primary lines of descent, in D. S. Bendall (ed.), *Evolution form molecules to man,* Cambridge University Press, London, pp. 209-233.

24. Cf. Glossary: "lipid bilayer", "taxonomy".

25. Woese, C. R. (1998) The universal ancestor, *Proc. Natl. Acad. Sci. USA* **95**, 6854-6859.

26. Prieur, D. Ercuso, G. and Jeanthon, C. (1995) Hyperthermophilic life at deep-sea hydrothermal vents, *Planet. Space Sci.* **43**, 115-122.

CHAPTER 5.

FROM THE AGE OF THE PROKARYOTES TO THE ORIGIN OF EUKARYOTES

1. Cf. Glossary: "natural selection", "genetic drift", "symbiosis"; cf. also Desmond, A. and Moore, J. (1991) *Darwin*. Michael Joseph, London, pp. 412-413.

2. Cf. Glossary: "red bed"; cf. also Han, T.-M. and Runnegar, B. (1992) Megascopic eukaryotic algae from the 2.1-billion-year-old Negaunee iron-formation, Michigan, *Science* **257**, 232-235.

3. Schopf, J. W. (1993) Microfossils of the Early Archean Apex Chert: New Evidence of the Antiquity of Life, *Science* **260**, 640-646.

4. Cf. Glossary: "Proterozoic"; cf. also Runnegar, B. (1992) Origin and Diversification of the Metazoa, in Schopf, J. W. and Klein, C. (1992) *The Proterozoic Biosphere,* Cambridge University Press, New York, p. 485.

5. Knoll, A. H. (1994) Proterozoic and Early Cambrian protists: Evidence for accelerating evolutionary tempo, *Proc. Natl. Acad. Sci. USA* **91**, 6743-6750.

6. Cf. Glossary: "Phanerozoic"; cf., also Conway-Morris, S. (1993) The fossil record and the early evolution of the Metazoa, *Nature* **361**, 219-225.

7. We refer the reader to the definitions of intron/exon in the Glossary and to Brenner, S. (1994) The ancient molecule, *Nature* **367**, 228-229.

8. Cf. Glossary: "progenote"; cf. also Maynard Smith, J. and Szathmary, E. (1993) The origin of chromosomes I. Selection for linkage, *J. Theor. Biol.* **164**, 437-446.

9. Cf. Glossary: "gene expression"; cf., also Maynard Smith, J. (1993) *The theory of evolution,* Canto Edition, Cambridge University Press, London, p. 122.

10. Chela-Flores, J. (1987) Towards a collective biology of the gene, *J. theor. Biol.* **126**, 127-136.

11. Chela-Flores, J. (1992) Towards the Molecular Basis of Polymerase Dynamics, *J. theor. Biol.* **154**, 519-539, and Erratum, *J. theor. Biol.* **157**, 269.

12. The reader is advised to read the entries in the Glossary for chromosome, which include histone, nucleosome and interphase; cf. also Å: Angstrom in "Abbreviations". Early research is due to Finch, J. T. and Klug, A. (1976) *Solenoidal model for superstructure in chromatin, Proc. Natl. Acad. Sci. USA* **73**, 1897-1901.

13. Cf. Glossary: for a definition of 'chromatin' and for a detailed explanation of the DNA/protein ratio, the appropriate reference is Watson, J. D., Hopkins, N. H., Roberts, J. W., Steitz, J. A., and Weiner, A. M. (1987) *Molecular Biology of the Gene,* (4th. ed.), The Benjamin / Cummings Publishing Co., Menlo Park, California, p. 682.

14. Darnell, J. E., Lodish, H. and Baltimore, D. (1990) *Molecular Cell Biology,* (2nd ed.), W. H. Freeman & Co., New York, p. 322.

15. Kornberg, A. and Baker, T. A.(1992) *DNA Replication,* (2nd ed.),   W. H. Freeman and Co., New York, p. 18, and pp. 508-510.

16. Zukerkandl, E. and Pauling, L. (1965) Molecules as documents of evolutionary history, *J. Theor. Biol.* **8**, 357-366.

17. Ballard, J. W O., Olsen, G. J., Faith, D. P., Odgers, W. A., Rowell, D. M., and Atkinson,   P. W. (1992) Evidence from 12S ribosomal RNA sequences that onychophorans are modified arthropods, *Science* **258**, 1345-1348.

18. Olsen, G. J. and Woese, C. R. (1993) Ribosomal RNA: a key to phylogeny, *The FASEB Journal* **7**, 113-123.

19. Margulis, L. and Guerrero, R. (1991) Kingdoms in turmoil, *New Scientist* 23 March, 46-50.

20. Woese, C. R., Kandler, O., and Wheelis, M. L. (1990) Towards a natural system of organisms. Proposal for the domains Archaea, Bacteria, and Eucarya, *Proc. Natl. Acad. Sci. USA* **87**, 4576-4579.

21. Whittaker, R. H. (1969) New concepts of kingdoms of organisms, *Science* **163**, 150- 160.

22. Douglas, S. E., Murphy, C. A., Spenser, D. F., and Gray, M. W. (1991) Cryptomonad algae are evolutionary chimaeras of two phylogenetically distinct unicellular eukaryotes, *Nature* **350**, 148-151.

23. Cavalier-Smith, T. (1987) Eukaryotes with no mitochondria, *Nature* **326**, 332-333.

24. Bonen, L and Doolittle, W. F. (1976) Partial sequences of 16S rRNA and the phylogeny of blue-green algae and chloroplasts, *Nature* **261**, 669-673.

25. John, P. and Whatley, F. R. (1975) *Paracoccus denitrificans* and the evolutionary origin of the mitochondrion, *Nature* **254,** 495-498.

26. Margulis, L. (1993) *Symbiosis in Cell Evolution,* W.H. Freeman & Co., San Francisco.

27. Seckbach, J. (1972) On the fine structure of the acidophilic hot-spring alga *Cyanidium caldarium*: a taxonomic approach, *Microbios* **5**, 133-142.

28. Hall, D. T., Strobel, D. F., Feldman, P. D., McGrath, M. A. and Weaver, H. A. (1995) Detection of an oxygen atmosphere on Jupiter's moon Europa, *Nature* **373**, 677-679.

29. Noll, K. S., Johnson, R. E., Lane, A. L., Domingue, D. and Weaver, H. A. (1996) Detection of ozone on Ganymede, *Science* **273**, 341-343.

30. Noll, K. S., Roush, T. L., Cruikshank, D. P., Johnson, R. E. and Pendleton, Y. J. (1997) Detection of ozone on Saturn's satellites Rhea and Dione, *Nature* **388**, 45-47.

31. McKay, C.P. (1996) Oxygen and the rapid evolution of life on Mars, in Chela-Flores, J. and Raulin, F. (eds.), (1996) *Chemical Evolution: Physics of the Origin and Evolution of Life,* Kluwer Academic Publishers, Dordrecht, pp. 177-184.

32. Cf. Glossary: "organelles", "nucleus (cell)", "chloroplast", "Paleozoic" and "Mesozoic"; cf. also Amabile-Cuevas, C. F. and Chicurel, M. E. (1993) Horizontal Gene Transfer, *American Scientist* **81**, 332-341.

33. For a recent discussion of the role of HGT in evolution, we refer the reader to Pennisi, E. (1999) Is it time to uproot the tree of life?, *Science* **284**, 1305-1307.

34. Chela-Flores, J. (1995) Molecular relics from chemical evolution and the origin of life in J. Chela-Flores, M. Chadha, A. Negron-Mendoza, and T. Oshima (eds.), *Chemical Evolution: Self-Organization of the Macromolecules of Life (A Cyril Ponnamperuma Festschrift),* A. Deepak Publishing, Hampton, Virginia, pp. 185-200.

35. De Duve, C. (1995) *Vital Dust. Life as a Cosmic Imperative,* Basic Books, New York, pp. 294-296.

36. Cf. Glossary: "chromomere".

37. Cf. Glossary: "contingency".

38. Cf. Glossary: "Orion Nebula".

CHAPTER 6.

EUKARYOGENESIS AND EVOLUTION OF INTELLIGENT BEHAVIOR

1. Teilhard de Chardin, P. (1954) *The Phenomenon of Man,* Collins, New York.

2. De Duve, C. (1995) *Vital dust: Life as a Cosmic Imperative,* Basic Books, New York.

3. Medawar, P. (1996) *The strange case of the spotted mice and other classic essays on science,* Oxford University Press, London.

4. Horvath, J., Carsey, F., Cutts, J., Jones, J., Johnson, E., Landry, B., Lane, L, Lynch, G., Chela-Flores, J., Jeng, T.-W. and Bradley, A. (1997) Searching for ice and ocean biogenic activity on Europa and Earth, in R.B. Hoover, (ed.), *Instruments, Methods and Missions for Investigation of Extraterrestrial Microorganisms,* The International Society for Optical Engineering, Bellingham, Washington USA, Proc. SPIE, **3111**, pp. 490-500.

5. In the Glossary the scientific significance of "chaos" is presented. The text refers to the work of Kauffman, S.A. (1993) *The origins of order: Self-Organization and Selection in Evolution,* Oxford University Press, London.

6. Davies, Paul (1998) Did Earthlife Come from Mars?) in Chela-Flores, J. and Raulin, F. (eds.), (1998) *Chemical Evolution: Exobiology. Matter, Energy, and Information in the Origin and Evolution of Life in the Universe,* Kluwer Academic Publishers, Dordrecht, pp.241-244.

7. De Duve, C. (1995) Loc. cit. pp. 294-296.

8. Farmer, J. (1992) Origins of multicellular individuality, in J. W. Schopf and C. Klein (eds.), *The Proterozoic Biosphere. A Multidisciplinary Study,* Cambridge University Press, New York, pp. 429-431.

9. Attenborough, D. (1979) *Life on Earth,* Fontana, London, p. 271.

10. van Eysinga, F. W. B.(1975) *Geological Time Table,* (3rd ed.), Elsevier Scientific Publishing Company, Amsterdam.

11. Chela-Flores, J. (1994) La vita nell'universo: verso una comprensione delle sue origini. In: Origini: l'universo, la vita, l'intelligenza, in F. Bertola, M. Calvani and U. Curi (eds.), Padova, Il Poligrafo, pp. 33-50.

12. Deacon, T. W. (1997) *The Symbolic Species. The Co-evolution of Language and Brain,* W.W. Norton & Company, New York.

13. Manger, P. Sum, M, Szymanski, M. Ridgway, S and Krubitzer, L. (1998) Modular Subdivisions of Dolphin Insular Cortex: Does Evolutionary History Repeat Itself?, *Journal of Cognitive Neuroscience* **10**, pp. 153-166.

14. Pinker, S. (1994) *The Language Instinct. How the Mind Creates Language,* William Morrow, New York.

15. Chomsky, Noam (1975) *Reflections on Language,* New York: Pantheon Books, pp. 10-11.

16. Dennett, D. (1995) *Darwin's Dangerous Idea Evolution and the Meanings of Life,* Penguin Books, London, pp. 384-393.

17. Dawkins, Richard (1983) *Universal Darwinism,* in Evolution form molecules to men, Bandell, D.S. (ed.), Cambridge University Press: London, pp. 403-425.

18. Medina-Callarotti, M. E. (2000) *Origins of Langiuage. The evolution of human speech,* in , in: Chela-Flores, J., Lemarchand, G.A. and Oro (eds.) *Astrobiology from the big bang to civilization.* Kluwer Academic Publishers: Dordrecht, The Netherlands. pp. 225-232.

CHAPTER 7.

ON THE POSSIBILITY OF BIOLOGICAL EVOLUTION ON MARS

1. Bonner, W. A. (1991) The origin and amplification of biomolecular chirality, *Origins of Life and the Evolution of the Biosphere* **21**, 59-111.

2. Carr, M. H. (1999) Mars: surface and interior, in P. R. Weissman, L.-A. McFadden and T. V. Johnson (eds.), *Encyclopedia of the Solar System,* Academic Press, San Diego, pp. 291-308.

3. Cf. Glossary: "pyrolysis", "gas chromatography" and "mass specrometer"; cf. also Soffen, G. A. (1976) Scientific results from the Viking Mission, *Science* **194**, 1274-1276.

4. McKay, C. P. and Stoker, C. R. (1989) The early environment and its evolution on Mars: Implications for Life, *Reviews of Geophysics* **27**, 189-214; McKay, C.P. (1996) Oxygen and the rapid evolution of life on Mars, in Chela-Flores, J. and Raulin, F. (eds.),. loc. cit., Preface, ref. 18, pp. 177-184.

5. Oró, J. (1996) Cosmic evolution, life and man, in Chela-Flores, J. and Raulin, F. (eds.). (1996) loc. cit., Introduction, ref. 18, pp. 3-19.

6. Hartman, H. and McKay, C. P. (1995) Oxygenic photosynthesis and the oxidation state of Mars, *Planet. Sci.* **43**, 123-128.

7. Kobayashi, K., Sato, T., Kaneko, T., Ishikawa, Y. and Saito, T. (1996) Possible formation of carbon compounds on Mars, in 11th International Conference on the Origin of Life, Orleans. Book of Abstracts, p. 61.

8. Farmer, C. B. and Doms, P. E. (1979) Global seasonal variation of water vapor on Mars, *J. Geophys. Res.* 84, 2881-2888.

9. The reader is referred to Table 1 in the following paper for references to the original literature: Sleep, N. H. (1994) *J. Geophys. Res.* **99**, 5639-5655.

10. Farmer, J. D. (1997) Implementing a strategy for Mars Exopaleontology, in R. B. Hoover, (ed.), *Instruments, Methods and Missions for Investigation of Extraterrestrial Microorganisms,* The International Society for Optical Engineering, Washington, Proc. SPIE **3111**, pp. 200-212.

11. Cf. Glossary: "plate tectonics", "magnetic anomaly patterns", "sea-floor spreading"; cf. also Kerr, R. A. (1999) Signs of plate tectonics on an infant Mars, *Science* **284**, 719-720; McKay, C. P. (1998) The search for extraterrestrial biochemistry in Mars and Europa, loc. cit. Introduction, ref. 19, pp. 219-228.

12. Haynes, R. H. and McKay, C. P. (1995) Designing a biosphere for Mars in Chela-Flores, J., M. Chadha, A. Negron-Mendoza, and T. Oshima (eds.), *Chemical Evolution:*

Self-Organization of the Macromolecules of Life, A. Deepak Publishing, Hampton, Virginia, pp. 201-211.

13. Head III, J.W., Hiesinger, H., Ivanov, M.A., Kreslavsky, M.A., Pratt, S.,and Thomson, B.J. (1999) Possible ancient oceans on Mars: Evidence from Mars Orbiter Laser Altimeter Data, *Science 286,* 2134-2137

14. Sims, M.R. *et al* (1999) Beagle 2 : a proposed exobiology lander for ESA's 2003 Mars Express Mission, *Adv. Space Res.* **23**, No.11 1925-1928.

15. Westall, F. *et al.* (2000) An ESA study for the search for life on Mars, *Planet. Space Sci.* 48 181-202.

16. White, R. J. and Averner, M. (2001) Humans in Space, *Nature* **409**, 1115-1118.

CHAPTER 8.

ON THE POSSIBILITY OF BIOLOGICAL EVOLUTION ON EUROPA

1. Rosino, L. (1992) *Exploration and discovery of the heavens* in F. Bertola (1992) *Da Galileo alle Stelle,* Biblos Edizioni, Padova.

2. Soffen, G. A. (1976) Scientific results from the Viking Mission, *Science* **194**, 1274-1276.

3. Horneck, G. (1995) Exobiology, the study of the origin, evolution and distribution of life within the context of cosmic evolution: a review, *Planet. Space Sci.* **43**, 189-217.

4. Greely, R. and Batson, R. (1997) *The NASA Atlas of the Solar System,* Cambridge University Press, London.

5. Oró, J. Squyres, S. W., Reynolds, R. T., and Mills, T. M. (1992) Europa: Prospects for an ocean and exobiological implications, in G. C. Carle, D. E. Schwartz and J. L. Huntington (eds.), *Exobiology in Solar System Exploration,* NASA SP 512, pp. 103-125.

6. Chela-Flores, J. (1998) First steps in eukaryogenesis: Origin and evolution of chromosome structure, *Origin Life Evol. Biosphere* **28**, 215-225.

7. Chela-Flores, J. (1996) Habitability of Europa: possible degree of evolution of Europan biota, *Europa Ocean Conference,* San Juan Capistrano Research Institute, San Juan Capistrano, California, USA, 12-14 November, 1996. p.21.

8. Shapiro, R. (1994) *L'origine de la vie,* Flammarion, Paris, pp. 402-404. For a more detailed account we refer to: Feinberg, G. and Shapiro, R. (1980) *Life beyond Earth: The intelligent Earthling's Guide to Life in the Universe,* William Morrow and Co,. New York, pp. 328-332.

9. Chela-Flores, J. (1997) A Search for Extraterrestrial Eukaryotes: Biological and Planetary Science Aspects, in C. B. Cosmovici, S. Bowyer and D. Werthimer (eds.), *Astronomical and Biochemical Origins and the Search for Life in the Universe,*. Editrice Compositore, Bologna, pp. 525-532.

10. Cf. Glossary: "tidal heating"; cf. also Reynolds, R. T., McKay, C. P. and Kasting, J.F. (1987) Europa, tidally heated oceans, and habitable zones around giant planets, *Adv. Space Res.* **7**, 125-132.

11. The comparison in table 8.4 is only an approximate correlation, which neglects several factors that will influence satellite formation; cf., also Reynolds, R. T., Squyres, S. W., Colburn, D. S. and McKay, C. P. (1983) On the habitability of Europa, *Icarus* **56**, 246-254.

12. Seckbach, J. (1972) On the fine structure of the acidophilic hot-spring alga *Cyanidium caldarium:* a taxonomic approach, *Microbios,* **5**, 133-142.

13. Petroni, G., Spring, S., Schleifer, K.-H., Verni, F., and Rosati, G. (2000) *Proc. Natl. Acad. Sci. USA* **97**, 1813-1817..

14. Rosati, G., Lenzi, P. and Franco, U. (1993) 'Epixenosomes': peculiar epibionts of the protozoon ciliate *Euplotidium itoi:* Do their cytoplasmic tubules consist of tubulin?, *Micron* **24**, 465-471.

15. Westall, F., Boni, L. and Guerzoni, E. (1995) The experimental silicification of microorganisms, *Paleontology* **38**, 495-528.

16 . Chela-Flores, J. (1998) A Search for Extraterrestrial Eukaryotes: Physical and Biochemical Aspects of Exobiology, *Origins Life Evol. Biosphere* **28,** 583-596.

17. Chela-Flores, J. (1994) Towards the theoretical bases of the folding of the 100-Å nucleosome filament, *J. theor. Biol.* **168**, 65-73.

18. Alberts, B., Bray, D., Lewis, J., Raff, M. Roberts, K. and Watson, J. D. (1989) *Molecular Biology of the Cell,* (2nd. ed.), Garland, New York.

19. Phillips, Cynthia B. and Chyba, Christopher F. (2001) Europa: Prospects for an Ocean and Life, in: Chela-Flores, J., Owen, Tobias and Raulin, F. (2001). *The First Steps of Life in the Universe.* Proceedings of the Sixth Trieste Conference on Chemical Evolution. Trieste, Italy, 18-22 September. Kluwer Academic Publishers: Dordrecht, The Netherlands. (In preparation.)

20. Joan Horvath, Frank Carsey, James Cutts, Jack Jones, Elizabeth Johnson, Bridget Landry, Lonne Lane, Gindi Lynch, Julian Chela-Flores, Tzyy-Wen Jeng and Albert Bradley (1997) Searching for ice and ocean biogenic activity on Europa and Earth, in: *Instruments, Methods and Missions for Investigation of Extraterrestrial Microorganisms,* The International Society for Optical Engineering, Bellingham, Washington USA. (R.B.Hoover, Ed.), Proc. SPIE, **3111**, pp. 490-500.

21. Cf. ref. 4, Chapter 5.

22. Delaney, J., Baross, J., Lilley, M. and Kelly, D. (1996) Hydrothermal systems and life in our solar system, *Europa Ocean Conference*, loc. cit., p. 26.

23. Cf. Glossary: "hydrothermal vent"; cf. also Huber, C. and Wachtershauser, G. (1988) Peptides by activation of amino acids with CO on (Ni, Fe)S surfaces: Implications for the origin of life, *Science* **281**, 670-672; cf. also Imai, E., Honda, H., Hatori, K, Brack, A. and Matsuno, K., (1999) Elongation of Oligopeptides in a Simulated Submarine Hydrothermal System, *Science* **283**, 831-833.

24. Chyba, C. (1998) Buried beginnings, *Nature* **395**, 329-330.

25. Anderson, R. T., Chapelle, F. H. and Lovly, D. R. (1998). Evidence against hydrogen-based microbial ecosystems in basalt aquifers, *Science* **281**, 976-977.

26. Stevens, T. O. and McKinley, J. P. (1995) Lithoautotrophic microbial ecosystems in deep basaltic aquifers, *Science* **270**, 450-454.

27. The Galileo Mission has shown us that the central part of some linae are of lower albedo than the surrounding terrain. Some planetary scientist believe that these bright surface features may represent fresh ice that has come from below. The darker parts of the linae may represent silicate contamination from below, or alternatively ice that may have been darkened by other external or internal factors; cf. Glossary, "linae".

28. Cf. Glossary: "tidal heating".

29.Cf. Glossary: "biochemistry", "lipid", "lipid bilayer"; cf. also Chapter 3, "The dawn of unicellular organisms".

30. Carr, M. H., Belton, M. J. S., Chapman, C. R., Davies, M. E., Geissler, P., Greenberg, R., McEwen, A S., Tufts, B. R., Greely, R., Sullivan, R., Head, J. W., Pappalardo, R. T., Klaasen, K. P., Johnson, T. V., Kaufman, J., Senske, D., Moore, J., Neukum, G., Schubert, G., Burns, J. A., Thomas, P. and Veverka, J. (1998) Evidence for a subsurface ocean in Europa, *Nature* **391**, 363-365.

31. Jakosky, B. (1998) *The Search for Life in Other Planets,* Cambridge University Press: London, p. 223.

CHAPTER 9.

ON THE POSSIBILITY OF CHEMICAL EVOLUTION ON TITAN

1. Coustenis, A. and Lorenz, R. D. (1999) Titan, in P. R. Weissman, L.-A. McFadden and T. V. Johnson (eds.), *Encyclopedia of the Solar System,* Academic Press, San Diego, pp. 377-404.

2. Owen, T., Gautier, D. and Raulin, F. (1992) Titan. In: Exobiology in Solar System Exploration, in G. C. Carle, D. E. Schwartz and J. L. Huntington (eds.), *Exobiology in Solar System Exploration,* NASA SP 512, pp. 126-143.

3. Raulin, F., Bruston, P., Coll, P., Coscia, D., Gazeau, M.-C., Guez, L. and de Vanssay, E. (1995) *Exobiology on Titan,* in C. Ponnamperuma and J. Chela-Flores (eds.), *Chemical Evolution: The Structure and Model of the First Cell,* Kluwer Academic Publishers, Dordrecht, pp. 39-53.

4. Morrrison, D. and Owen, T. (1996) *The Planetary System,* (2nd ed.), Addison-Wesley Publishing Co., Reading, Mass.

5. The Cassini Program Pocket Reference (October 1997) Printed in USA, California Institute of Technology.

6. Jakosky, B. (1998) *The Search for Life in Other Planets,* Cambridge University Press: London.

7. Dermott, S.F. and Sagan, C. (1995) Tidal effects of disconnected hydrocarbon seas on Titan, *Nature* **374**, 238-240.

8. *ESA Huygens. Exploring Titan, a mysterious world.* An ESA publication.

9. Taylor, F. W. and Coustenis, A. (1998) Titan in the Solar System, *Planet. Space Sci.* **46**, 1085-1097.

10. McEwen, A. S., Kezsthelyi, L., Spenser, J. R., Schubert, G., Matson, D.L., Lopes-Gautier, R., Klaasen, K. P., Johnson, T. V., Head, J. W., Geissler, P., Fagents, S., Davies, A. G., Carr, M. H., Breneman, H. H. and Belton, M. J. S. (1998). High-Temperature Silicate Volcanism on Jupiter's Moon Io, *Science* **281**, 87-90.

11. Whitman, W.B., Coleman, D.C. and Wiebe, W.J. (1998) *Prokaryotes the unseen majority,* Proc. Natl. Acad. Sci USA **95,** 6578-6583.

CHAPTER 10.

HOW DIFFERENT WOULD LIFE BE ELSEWHERE?

1. Kauffman, S. A. (1993) *The origins of order: Self-Organization and Selection in Evolution,* Oxford University Press, London.

2. Davies, P. C. W. (1998) Did Earthlife come from Mars?, in Chela-Flores, J. and Raulin, F. (eds.). (1998) *Chemical Evolution: Exobiology: Matter, Energy, and Information in the Origin and Evolution of Life in the Universe,* Kluwer Academic Publishers, Dordrecht, pp. 241-244.

3. De Duve, C. (1995) *Vital dust: Life as a cosmic imperative,* Basic Books, 1995, p. 297.

4. Mayor, M. and Queloz, D. (1995) A Jupiter-mass companion to a solar-type star, *Nature* **378**, 355-359.

5. Sagan, C. (1980) *Cosmos,* Random House, New York, p. 209.

6. Mayor, M., Queloz, D., Udry, S. and Halbwachs, J.-L. (1997) From brown dwarfs to planets, in C. B. Cosmovici, S. Bowyer and D. Werthimer (eds.), *Astronomical and Biochemical Origins and the Search for Life in the Universe,* Editrice Compositore, Bologna, pp. 313-330.

7. Black, David C. (1999) *Extrasolar planets: Searching for other planetary systems.* In: "Encyclopaedia of the Solar System". P. R. Weissman, L.-A. McFadden and T. V. Johnson (eds.), Academic Press, San Diego, pp. 377-404.

8. Marcy, G. (1998) Back in focus, *Nature* **391**, 127.

9. Lissauer, J. J. (1999) Three planets for Upsilon Andromedae, *Nature* **398**, 659-660.

10. Jakosky, B. (1998) *The Search for Life in Other Planets,* Cambridge University Press: London.

11. Cf. Glossary: "astrometry".

12. Cf. Glossary: "brown dwarf".

13. Cf. Glossary: "zodiacal dust cloud".

14. Cf. Glossary: "Orion Nebula".

CHAPTER 11.

THE SEARCH FOR THE EVOLUTION OF INTELLIGENT BEHAVIOR IN OTHER WORLDS

1. Cocconi, G. and Morrison, P. (1959) Searching for interstellar communications, *Nature* **184**, 844-846.

2. Drake, F. and Sobel, D. (1992) *Is there anyone out there? The scientific search for Extraterrestrial Intelligence.* Delacorte Press, New York.

3. Zuckerman, B. and Hart, M. H. (1995) *Extraterrestrials. Where are they?,* (2nd ed.), Cambridge University Press, London.

4. Mayor, M. (1996) The Geneva radial velocity survey for planets, in C.B. Cosmovici, S. Bowyer and D. Werthimer (eds.), *Astronomical and Biochemical Origins and the Search for Life in the Universe,* Editrice Compositore, Bologna.

5. Tarter, J. (1998) The search for intelligent life in the universe, *Commentarii* (Vatican City) **4**, N.3., 305-310.

6. Lemarchand, G. A. (2000) Detectability of intelligent life in the universe, in Chela-Flores, J., Lemarchand, G.A. and Oro (eds.), J. (2000). *Astrobiology: Origins from the Big Bang to Civilisation,* Kluwer Academic Publishers, Dordrecht, The Netherlands, pp. 13-32.

7. Oro, J. (1998) The Abdus Salam Lecture, in Chela-Flores, J. and Raulin, F. (eds.), loc. cit "Introduction", ref. 19, pp. 11-32.

8. Heidmann, J. (1996) SETI from the Moon. A case for a XX Century SETI-Dedicated Lunar Farside Crater, in Chela-Flores, J. and Raulin, F. (eds.), (1996) loc. cit. Preface, ref. 18, pp. 343-353.

9. Genta, G. (2001) The Saha Crater Radioastronomic and SETI Obsservatory, in Chela-Flores, J., Owen, Tobias and Raulin, F. *The First Steps of Life in the Universe,* Kluwer Academic Publishers, Dordrecht, in preparation.

10. Goldsmith, D. (1999) New telescope will turn a keen ear on E.T,*Science* **283**, 914.

11. Horowitz, P., cited in: *Is Anybody Out There?* Time Magazine February 5, 1996. pp. 42-50.

12. Cf. Glossary: "cell signaling" and "tissues" and "mitosis".

13. Cf. Glossary: "lipid bilayer". and Chapter 16 (Coda).

14. Drake, F. and Sobel, S., loc. cit., pp. 45-64.

15. Sagan, C. (1973) Discussion, in C. Sagan (ed.), *Communication with Extraterrestrial Intelligence (CETI),* The MIT Press, Cambridge, Mass., pp. 112-146.

16. Cf. Glossary: "geocentricism", "anthropocentrism", and biogeocentrism".

CHAPTER 12.

IS THE EVOLUTION OF INTELLIGENT BEHAVIOR UNIVERSAL?

1. Wilson, Edward O. (1998) *Consilience The unity of knowledge.* Alfred A. Knopf: New York, p. 70.

2. Drake, Frank, (2001) *New Paradigms for SETI,* in Chela-Flores, J., Owen, T. and Raulin, F. (2001). *The First Steps of Life in the Universe.* Proceedings of the Sixth

Trieste Conference on Chemical Evolution. Trieste, Italy, 18-22 September, 2000. Kluwer Academic Publishers: Dordrecht, The Netherlands. (In preparation.)

3. Villegas, R, Castillo, C. and Villegas, G.M. (2000) *The origin of the neuron: The first neuron in the phylogenetic tree of life,* in: Chela-Flores, J., Lemarchand, G.A. and Oro (eds.) *Astrobiology from the big bang to civilization.* Kluwer Academic Publishers: Dordrecht, The Netherlands. pp. 195-211.

4. Marino, L. (2000) *Turning the empirical corner on F : The probability of complex intelligence,* in *A New Era in Astronomy,* Lemarchand G.A. and Meech K. (eds.), ASP Conference Series, San Francisco, USA, **213,** 431-435.

5. Marino, L. (1997) *Brain-behavior relations in primates and cetaceans: Implications for the ubiquity of factors leading to the evolution of complex intelligence,* in: *Astronomical and Biochemical Origins and the Search for Life in the Universe.* C.B. Cosmovici, S. Bowyer and D. Werthimer (eds.). Editrice Compositore: Bologna. pp. 553-560.

6. Attenborough, D. (1995) *The Life of Plants.* BBC Books: London.

7. De Duve, C. (1995) *Vital Dust Life as a cosmic imperative,* Basic Books, New York.

CHAPTER 13.

DEEPER IMPLICATIONS OF THE SEARCH FOR EXTRATERRESTRIAL LIFE

1. Russell, B. (1991) *History of Western Philosophy and its Connection with Political and Social Circumstances from the Earliest Times to the Present Day,* Routledge, London, p.13.

2. Gould, S. J. (1991) *Wonderful life. The Burgess Shale and the Nature of History.* Penguin Books, London, pp. 48-52.

3. Conway-Morris, S. (1998) *The Curcible of Creation. The Burgess Shale and the Rise of Animals,* Oxford University Press: New York, pp. 9-14.

4. Tucker Abbott, R. (1989) *Compendium of landshells,* American Malacologists, Melbourne, Florida, USA, pp. 7-8; Austin, Jr., A.L. (1961) *Birds of the World,* Paul Hamlyn, London, p. 216.

5. Drake, F. and Sobel, D. (1992) *Is there anyone out there? The scientific search for Extraterrestrial Intelligence,* Delacorte Press, New York, pp. 45-64.

6. Mayor, M., Queloz, D., Udry, S. and Halbwachs, J.-L. (1997) From Brown Dwarfs to planets, in C.B. Cosmovici, S. Bowyer and D. Werthimer (eds.), *Astronomical and Biochemical Origins and the Search for Life in the Universe,* Editrice Compositore, Bologna, pp. 313-330.

7. Teilhard de Chardin, P. (1965) *The phenomenon of man,* Fontana Books, London, p. 33.

8. Chela-Flores, Julian (1997) Cosmological models and appearance of intelligent life on Earth: The phenomenon of the eukaryotic cell, in Padre Eligio, G. Giorello, G. Rigamonti and E. Sindoni (eds.), *Reflections on the birth of the Universe: Science, Philosophy and Theology,* Edizioni New Press, Como, pp. 337-373.

9. Chela-Flores, J. (1998) The Phenomenon of the Eukaryotic Cell, in R. J. Russell, W. R. Stoeger, and F. J. Ayala (eds), *Evolutionary and Molecular Biology: Scientific Perspectives on Divine Action,* Vatican City State/Berkeley, California and Vatican Observatory and the Center for Theology and the Natural Sciences (a joint publication), pp. 79-99.

10. De Duve, Christian (1995)*Vital dust: Life as a cosmic imperative,* Basic Books, New York, pp. 160-168.

11. John Paul II (1996) Papal Message to the Pontifical Academy, *Commentarii* 4, N. 3. Vatican City, 1997, pp. 15-20; cf. also: La traduzione in italiano del Messaggio del Santo Padre alla Pontificia Accademia delle Scienze, *L'Osservatore Romano* 24 October 1996, p. 7.

12. The question of the impact of extraterrestrial life on our culture, a discipline referred to as "astrotheology", has been reviewed in Dick, S. J. (1998) *Life on Other Worlds,* Cambridge University Press, London, pp. 245-256.

13. Simpson, W. K. (ed.), (1972) *The Literature of Ancient Egypt. An Anthology of Stories, Instructions and Poetry,* Yale University Press, New Haven, pp. 7-9 and pp. 289-295.

14. Cf. Glossary: "natural theology"; cf., also Russell, J. R. (1995) Introduction in *Quantum Cosmology and the Laws of Nature. Scientific Perspectives on Divine Action,* 2nd edition. R. J. Russell, N. Murphy and C.J. Isham (eds.), Vatican Observatory Foundation, Vatican City State, pp. 1-31.

15. Barbour, I. G. (1995) Ways of relating science and theology, in R. J. Russell, W. R. Stoeger, SJ and G. V. Coyne, SJ (eds.), *Physics, Philosophy and Theology. A common quest for understanding,* 2nd. ed., Vatican Observatory Foundation, Vatican City State, pp. 21-48.

16. Coyne, SJ, G. V. (1998) *The concept of matter and materialism in the origin and evolution of life* in Chela-Flores, J. and Raulin, F. (Eds.), *Chemical Evolution: Exobiology. Matter, Energy, and Information in the Origin and Evolution of Life in the Universe,* Kluwer Academic Publishers, Dordrecht, pp. 71-80

17. Carroll, S.B. (2001) Chance and necessity: the evolution of morphological complexity and diversity, *Nature* 409, 1102-1109.

18. Krubitzer, L. (1995). The organization of neocortex in mammals: are species differences really so different? *Trends in Neuroscience* **18**, 408-417.

19. Peacocke, A. (1988) Biological evolution - a positive theological appraisal, in *Evolutionary and Molecular Biology: Scientific Perspectives on Divine Action.* R. J. Russell, W. R. Stoeger and F. J. Ayala, Editors. Vatican City State/Berkeley, California: Vatican Observatory and the Center for Theology and the Natural Sciences, pp. 357-376.

CHAPTER 14.

PHILOSOPHICAL   IMPLICATIONS   OF   THE   SEARCH   FOR EXTRATERRESTRIAL CIVILIZATIONS

1. Holmes, S. J. (1961) *Louis Pasteur,* New York, Dover, p. 51.

2. Flew, A. (ed.) (1979) *A dictionary of philosophy,* Pan Books, London, pp. 300-301.

3. Russell, R.J. (2001) Life in the universe: Philosophical and Theological Issues, in Chela-Flores, J., Owen, T. and Raulin, F., *The First Steps of Life in the Universe.* Proceedings of the Sixth Trieste Conference on Chemical Evolution. Trieste, Italy, 18-22 September. Kluwer Academic Publishers: Dordrecht, The Netherlands. (To be published.)

4. Ayala, F.J. (1998) Darwin's Devolution: Design without Designer, in *Evolutionary and Molecular Biology: Scientific Perspectives on Divine Action,* Robert John Russell, William R. Stoeger, S.J. and Francisco J. Ayala, Editors, Vatican City State/Berkeley, California: Vatican Observatory and the Center for Theology and the Natural Sciences. pp. 101-116.

5. Cf. Glossary: "revelation"; cf. also Russell, B. (1991) *History of Western Philosophy and its Connection with Political and Social Circumstances from the Earliest Times to the Present Day,* Routledge, London. p.13.

6. Townes, C. H. (1995) *Making Waves,* AIP, Woodbury, NY. pp. 157-167.

7. John Paul II (1992) Discorso di Giovanni Paolo II alla Pontificia Accademia delle Scienze. *L'Osservatore Romano,* 1st November, p. 8.

8. St. Augustine (1984) *Concerning the City of God against the Pagans,* Penguin Classics, London (cf. Book XVII, p.4).

9. Cf. Glossary: "revelation"; cf. also John Paul II (1996) Papal Message to the Pontifical Academy of Sciences of 22 October 1996, *L'Osservatore Romano Weekly Edition.* N. 44, 30 October, p.3 and p. 7. (A translation from the official version in the French Language.)

10. Polkinghorne, John (1996) *Scientists as theologians,* SPCK, London.

11. Olavi-Kajander, E. and Çiftçioglu, N. (1998) Nanobacteria: An alternative mechanism for pathogenic intra- and extracellular calcification and stone formation, *Proc. Natl. Acad. Sci. USA* **95**, 8274-8279.

12. Carson, D. A. (1998) An infectious origin of extra skeletal calcification, *Proc. Natl. Acad. Sci. USA* **95**, 7846-7847.

13. Humphries, C. J. , Press, J. R. and Sutton, D. A. (1989) *The Hamlyn Press to trees of Britain and Europe,* Hamlyn, London, pp. 50-51.

14. By biogeocentrism we mean the belief that life has occurred only on Earth. (cf. the corresponding definition in the Glossary.)

15. Mayr, E. (1995) The search for extraterrestrial intelligence *in Extraterrestrials. Where are they?,* in B. Zuckerman and M.H. Hart (eds.), (2nd. ed.), Cambridge University Press, London, pp. 152-156.

16. Monod, J. (1972) *Chance and Necessity An Essay on the Natural Philosophy of Modern Biology,* Collins, London, p. 136.

17. Conway-Morris, S. (1998) *The Curcible of Creation. The Burgess Shale and the Rise of Animals,* Oxford University Press: New York, pp. 222-223.

18. Rilke, R.M. (1998) *Elegie Duinese,* Biblioteca Universale Rizzoli: Milan.

19. Fischer, E.P. and Lipson, C. (1988) *Thinking about Science Max Delbrück and the Origins of Molecular Biology,* W.W. Norton: New York, pp. 288-291.

20. Russell, B. (1997) *Religion and Science,* Oxford University Press, New York, pp. 49-81.

21. Burrows, W. E. (1998) *This New Ocean: The Story of the First Space Age,* Random House, London.

22. Jastrow, R. (1997) The place of humanity in the cosmic community of intelligent beings, in Hoover, R. B., *Instruments, Methods and Missions for Investigation of Extraterrestrial Microorganisms,* The International Society for Optical Engineering, Bellingham, WA, USA Proc. SPIE, **3111**, pp.15-23.

CHAPTER 15.

BACK TO THE BEGINNING OF ASTROBIOLOGY

1. Bertola, F. (1992) Seven centuries of astronomy in Padua, in *From Galileo to the Stars,* Biblos Edizione, Cittadella (PD), Italy, pp.71-74.

2. Russell, B. (1961) *History of Western Philosophy,* G. Allen & Unwin, London

3. Gatti, H. (1999) *Giordano Bruno and Renaissance Science,* Cornell University Press, Ithaca.

4. Keynes, R.D. (1998) The Theory of Common Descent, in Chela-Flores, J. and Raulin, F. (eds.), *Exobiology: Matter, Energy, and Information in the Origin and Evolution of Life in the Universe,* Kluwer Academic Publishers, Dordrecht, The Netherlands, pp. 35-49.

5. Schopf, J. W. (1999) *Cradle of Life the discovery of Earth's earliest fossils,* Princeton University Press: Princeton, p. 112.

6. Ponnamperuma, C. (1995) The origin of the cell from Oparin to the present day, Ponnamperuma, in C. and Chela-Flores, J. (eds.), *Chemical Evolution: The Structure and Model of the First Cell,.* Kluwer Academic Publishers, Dordrecht, The Netherlands, pp.3-9.

7. Chadha, M. (2001) Private communication. For an extensive pictorial record of recent events in the history of astrobiology we refer the reader to the following web site: http://www.geocities.com/issol_trieste_in_pictures/sld092.htm

CHAPTER 16.

RECAPITULATION

1. Chela-Flores, J. (1997) Cosmological models and appearance of intelligent life on Earth: The phenomenon of the eukaryotic cell, in: "Reflections on the birth of the Universe:Science, Philosophy and Theology". Eds. Padre Eligio, G. Giorello, G. Rigamonti and E. Sindoni. Edizioni New Press: Como, 1997. pp. 337-373.

2. Chela-Flores, J. (1998) The Phenomenon of the Eukaryotic Cell, in: *Evolutionary and Molecular Biology: Scientific Perspectives on Divine Action.* R. J. Russell, W. R. Stoeger and F. J. Ayala, Editors. Vatican City State/Berkeley, California: Vatican Observatory and the Center for Theology and the Natural Sciences, pp. 79-99.

Glossary

(Words in italics in the definition of a given term are defined elsewhere in the Glossary)

Accretion. In the *solar nebula* it is the early stage of accumulation of mass into a protoplanet.

Albedo. Percentage of incoming visible radiation reflected by the surface.

Alcohol. A derivative of *hydrocarbons* containing the chemical group OH (a 'hydroxyl') in place of a hydrogen atom.

Alga. Any of a group of mainly aquatic organisms that contain chlorophyll and are able to carry out *photosynthesis.*

Amino acid. Any of the twenty organic compounds that are the building blocks of proteins, when they are synthesized at *ribosomes,* according to the rules dictated by the *genetic code.* Their precise chemical formula is not essential for following the arguments in this book. They are listed in Chapter 3 in the section on "The Genetic Code".

Angular momentum. A property of rotating objects expressed as mvr, where m denotes mass, v denotes velocity and r is the distance from the center of rotation.

Anthropocentrism. Doctrine that maintains that man is the center of everything, the ultimate end of nature.

Archaea. cf., archaebacteria.

Archaebacteria. These are a group of single-celled organisms which are neither bacteria nor *eukaryotes.* They may be adapted to extreme conditions of temperature (up to just over 100 ° C), in which case they may be called thermophiles. They may also be adapted to extreme acidic conditions (acidophiles). The alternative expression of *extremophiles* is used sometimes for these organisms to distinguish different degrees of adaptability to such extreme ranges of conditions. All archaebacteria are said to form the domain *Archaea,* divided into kingdoms.

Archean. In geologic time, this is an era that spans from 4.5 to 2.5 Gyr BP. (The Hadean is the first of its suberas). We refer to this era together with the *Proterozoic* (2.5-0.57 Gyr BP) as the *Precambrian Eon.* (The eon in which multicellular eukaryotic life arose is called the *Phanerozoic* and it ranges from the end of the Precambrian till the present).

Arnothosite. A deep-seated rock found on Earth and retrieved with the Moon samples of the *Apollo missions.* It is formed from a solidified form of *magma.*

215

Astrobiology. This term stands for the science of the origin, evolution, distribution and destiny of life in the universe. Astrobiology is currently in a period of fast development due to the many space missions that are in their planning stages, or indeed are already in operation. (Cf., *bioastronomy* and *exobiology.*)

Astrometry. This subject concerns the observation of the position of celestial objects and its variation over time.

Recently, astrometry has been applied to the measurable wobbling motion of a star, due to its rotating suite of planets, against a fixed background of stars. This phenomenon has been used to discover the existence of extrasolar planets.

Astronomical unit. The average distance between the Sun and the Earth, approximately 150 million kilometers, or equivalently 93 million miles. It is abbreviated as AU.

Banded iron formation. These are layered compounds of silica and iron. The layers may contain other sedimentary rock, notably chert.

Basalt. Common *rock* that has solidified from exposed *magma,* composed mainly of silicon, oxygen, iron, aluminum and magnesium. It is abundant in the oceanic crust, and in general on the surface of the *terrestrial planets.*

Base. *Purines* or *pyramidines* in *DNA* or *RNA.*

Bioastronomy. A synonym of *exobiology.* It emphasizes research with the specific tools of astronomy, especially *astrometry* and astronomy in the microwave, radio and visible regions of the electromagnetic spectrum. In view of the duality of the terms 'bioastronomy' and 'exobiology', for simplicity we use *'Astrobiology'* throughout the present book to encompass both disciplines. The terms bioastronomy and exobiology are reserved for the topics included in Parts I and II of Book 3.

Biochemistry. The study of the chemistry of living organisms. (It overlaps to a certain extent with *molecular biology.)*

Biogeocentrism. A term introduced in the text to reflect a tendency observed in some contemporary scientists and philosophers according to which life is only likely to have occurred on Earth.

Biomarker. A characteristic biochemical substance or *mineral* that can be taken to be a reliable indicator of the biological origin of a given sample.

Biota. The totality of life on Earth.

Breccia. *Rock* composed of fragments derived from previous generations of rocks.

Brown dwarf. A very cool star incapable of igniting nuclear reactions in its core. Gliese 229B, the first detected brown dwarf, was discovered in 1995; it has surface temperature of about 650°C and mass equivalent to about 50 Jupiter masses (M_J).

Like other brown dwarfs, below some 80 M_J the pressure is not high enough for nuclear fusion to occur (cf., "Origin of the elements", Chapter 1).

Theoretical modeling suggests that a lower bound for the formation of a brown dwarf is about 10 M_J.

C-type asteroid. A term referring to dark, carbonaceous asteroids in a classification according to the spectra of reflected sunlight.

Carbonaceous chondrite. A meteorite formed in the early solar system. Its constituents are *silicates,* bound water, carbon and organic compounds, including *amino acids.*

Carbonate. A mineral that releases carbon dioxide by heating.

Cell division. Separation of a cell into two daughter cells. In *eukaryotes* the *nucleus* divides as well *(mitosis).* This is followed by a division of the extranuclear contents inside the *lipid bilayer* of the plasma membrane.

Cell signaling. Communication between cells by extracellular chemical signals.

Cenancestor. (cf., *progenote).*

Cenozoic. The most recent era of the *Phanerozoic eon.*

Chaos. Chaotic systems display a level of behavioral complexity which frequently cannot be deduced from a knowledge of the behavior of their parts.

Chirality. The property of molecules that exist in two forms whose spatial configurations are mirror images of each other. An example repeatedly mentioned in the text are the protein *amino acids.*

Chloroplast. Any of the chlorophyll-containing organelles that are found in large numbers in photosynthesizing plant cells. (They are the sites of *photosynthesis.*) Chloroplasts are widely believed to be the relic of a once free-living *cyanobacterium.*

Chondrules. A millimeter-sized spheroidal particle present in some kinds of meteorites. Originally they were molten or partially molten droplets.

Chromatin. The structure of *eukaryotic chromosomes*, mainly DNA and protein.

Chromomere. A tight mass of DNA.

Chromosome. A cell structure which contains *DNA* and whose number and complexity depends on the degree of evolution of the cell itself. Typically its constituent DNA in eukaryotes folds around proteins called *histones,* forming a set of compact structures *'nucleosomes'* ; each structure is separated from others by a length of 'linker' DNA. The degree of folding are referred to by means of the dimension of the nucleosomes (in Ångstroms Å).

Codon. A triplet of *bases* in the *nucleic acid DNA* and also *RNA* that takes part in the process of *protein* synthesis) that codes for a given *amino acid.*

Color index of a star. These parameters are formed by taking the differences between the magnitude of a star measured in one wavelength band and its magnitude in another.

Contingency. The notion that the world today is the result of chance events in the past.

Convergent evolution. Independent evolution of similar genetic or morphological features.

Cyanobacterium. Prokaryote capable of oxygenic *photosynthesis.* Some cyanobacteria appear in the fossil record in the *Archean.* and are regarded to be ancestral to *chloroplasts.*

Diploblasts. Animals arising from an embryo consisting of two cell layers.

DNA (Deoxyribose nucleic acid). A substance present in every cell, bearing its hereditary characteristics.

Domain. In *taxonomy* it refers to the highest grouping of organisms, which include kingdoms and lower taxons, such as phyla or divisions, orders, families, genera and species.

Element. A substance that is irreducible to simpler substances in the sense that all its atoms have the same number of protons and electrons, but may have a different number of neutrons (cf., also *'isotope'*).

Enantiomer. One of the two forms of mirror images of chiral compounds.

Euchromatin. Loosely-packed *chromatin* (cf., *heterochromatin).*

Eukaryogenesis. The first appearance of the *eukaryotes,* usually believed to be partly due to the process of *symbiosis,* probably during the Late *Archean* 2.7 billion years before the present (Gyr BP), according to current views; but certainly during the *Proterozoic,* some 1.8 Gyr BP, eukaryotes were co-existing with *prokaryotes,* both bacteria and *archaebacteria.*

Eukaryotes. These are either single-celled, or multicellular organisms in which the genetic material is enclosed inside a double membrane, which is called nuclear envelope. Taxonomically the totality of such organisms is said to form the domain *Eucarya,* which contains kingdoms, such as Animalia.

Exobiology. The study of extraterrestrial life based on astronomy, physics and chemistry, as well as the earth and life sciences (cf., *astrobiology* and *bioastronomy).*

Exon. A gene sequence coding for part or the whole of a gene product.

Extremophiles. This term is explained under *"Archaebacteria"*.

Faith. Subjective response to Divine truth; supernatural act of the will.

Fluorescence. A physical property of some substances, which consists of being able to absorb light at a given frequency and re-emitting it at a longer wavelength. *DNA* and *RNA* are able to bind certain dyes and are thereby subject to detection by means of fluorescent microscopy.

Gas chromatography. A chemical technique for separating gas mixtures, in which a gas goes through a column containing an absorbent phase that separates the gas mixture into its components.

Gene expression. The process that uses *RNA* by means of which a gene translates the information it codifies, according to the rules of the *genetic code,* into *proteins.*

Genetic code. This is a small dictionary which relates the four letter language of *nucleic acids* to the twenty-letter language of the *proteins* (cf., Table 4.3).

Genetic drift. Evolutionary change in small populations produced by random effects, not by *natural selection.*

Genome. The set of all genes contained in a single set of *chromosomes* of one species.

Geocentric. An old hypothesis that maintained that the Earth was at the center of the universe; in particular it maintained that the Sun was in orbit around the Earth.

Geochronology. The application and interpretation of isotopic dating methods in geology.

Glycerol. A colorless, sweet-tasting viscous liquid, widely distributed as part of molecules of all living organisms. (At the molecular level its atomic formula is that of an *alcohol.*)

Hadean. The earliest subera of the *Archean.*

Hematite. A *mineral* of ferric oxide.

Hertzprung-Russell diagram. This useful plot of luminosity of stars versus surface temperature (or equivalent physical parameters) was introduced independently by Ejnar Hertzprung and Henry Russell in the early part of the 20th century. At various stages of *stellar evolution,* stars will occupy different parts on this diagram.

Heterochromatin. A densely packed form of *chromatin* in eukaryotic chromosomes.

Histone. A protein belonging to one of the four classes H2A, H2B, H3 and H4. A couple of representatives of each class, together with a stretch of double-helix DNA, forms a nucleosome.

Hominid (Hominidae). The name given to the taxon which comprises man and his ancestors, such as the Australopithecine.

Hydrocarbon. A class of organic compounds in which only hydrogen and carbon take part.

Hydrothermal vent. These structures occur at the crest of oceanic ridges, producing the ascension of a very hot suspension of small particles of solid *sulfide* in water (cf., *plate tectonics, sea-floor spreading).*

Interphase. The period following the completion of *cell division* when its *nucleus* is not dividing.

Interstellar dust. The ensemble of solid particles borne by the gas that occupies the space between the stars.

Intron. A typically eukaryotic sequence in a gene that does not code for amino acids of a protein.

Ion. An atom or molecule that is electrically charged due to the loss or gain of one or more electrons.

Isotope. One of two or more atoms of the same chemical *element* that have the same number of protons in their *nucleus*, but differ in their number of neutrons.

Jovian planet. Collective name for Jupiter, Saturn, Uranus and Neptune. These are all giant planets, hence very different form the *terrestrial planets.*

Kuiper belt. A belt of some billion (10^9) or more comets beyond the orbit of Neptune. It is the source of short-period comets. It extends possibly as far as 1000 *astronomical units.*

Lava. Molten *rock* material that is expelled from volcanoes. It consolidates once it reaches the surface of the Earth or the sea-floor.

Light-year. The distance light travels in a year, approximately ten trillion miles (10^{13} Km).

Linea. Elongate markings first observed on the surface of Europa by the *Voyager mission,* which are a few kilometers wide and of lower albedo than the surrounding terrain.

Lipid.   A wide group of organic compounds having in common their solubility in organic solvents, such as *alcohol.* They are important in biology, as they are constituents of the cell *membrane* and have a multitude of other important roles.

Lipid bilayer.   A characteristic fabric of all biological *membranes*. It consists of two monolayers of *lipids.*

Logical positivism.   A philosophical current that maintains that scientific knowledge is the only kind of factual knowledge; all traditional doctrines are to be rejected as meaningless. These ideas were mainly developed in the 1920s in Vienna.

Magma. Liquid or molten *rock* material, called *lava* when it reaches the Earth's surface.

Magnetic anomaly patterns   (also called 'magnetic stripes' or 'magnetic banding'). A phenomenon that has been observed both on Earth and Mars. Although the origin of the Martian magnetic stripes is unclear at the time of writing (cf., Chapter 7, Table 7.6), the origin on the Earth is clear:
It is due to the *sea-floor spreading*, which produces new crust very slowly. Consequently, as the polarity of the Earth magnetic field is reversing in a time scale comparable with the creation of new crust, this results in different magnetic signatures that can be read in 'bands' of sea-floor, each of which has a characteristic magnetic polarity typical of the time when the corresponding new crust was formed.

Mass spectrometer.   Instrument used for the separation of a beam of gaseous *ions* into components with different values of mass divided by charge. The ion beams are detected with an electrometer.

Maria.   It is the plural of mare. Latin for sea. It refers to the dark patches on the surface of the Moon; the darkness arises from *basaltic lava* flows.

Membrane.   A tissue consisting of *lipids, protein* and sugars (polysaccharides) that covers biological cells and some of their organelles. The cell nucleus is covered by an envelope made of two membranes.

Mesozoic.   An era of the *Phanerozoic eon,* which was characterized by reptiles together with brachiopods, gastropods and corals.
      The best known period of this era is the Jurassic period (208-146 Myr BP) with rich flora and warm climate when reptiles, mainly dinosaurs, were dominant on land.

Messenger RNA (mRNA).   An *RNA* molecule that specifies the *amino acid* sequence of a protein.

Metaphase. The second stage of *cell division,* during which the envelope around the *nucleus* breaks down in some eukaryotes such as animals.

Mitochondrion. A eukaryotic *organelle* that is the main site of energy production in most cells.

Mitosis. Division of the *nucleus* of the *eukaryotes*. During this process the DNA is condensed into visible *chromosomes*.

Molecular biology. The study of the structure and function of the macromolecules associated with living organisms.

Natural selection. Darwin and Wallace suggested this term for the reproductive success. It refers to the selection which eliminates most organisms that over reproduce, and hence allows adaptation to changing environmental conditions.

Natural theology. The body of knowledge about religion which can be obtained by human *reason* alone, without appealing to *revelation*.

Neutron star. A relic of a *supernova* explosion. It consists of subnuclear particles called neutrons in various states of condensed matter.

Nitrile. A class of chemical compound that contains the group CN, which on hydrolysis yields ammonia and a carboxylic acid.

Nucleic acid. These are organic acids whose molecular structure consists of five-carbon sugars, a phosphate and one of the following five *bases:* adenine, guanine, uracil, thymine and cytosine (sometimes modified).

Nucleosome. A bead-like subunit of *chromatin* in the *chromosomes* of *eukaryotes*. It consists of a length of *DNA* double-helix wrapped around a core of eight *histones*.

Nucleosynthesis. The generation of the natural *elements* starting from hydrogen. Three possibilities are: at an early stage of the big bang, after temperatures had cooled down sufficiently to allow the electromagnetic force to produce atoms; at the center of stars; and during the *supernova* stage of *stellar evolution*.

Nucleus of an atom (plural: nuclei). Massive central body composed of protons neutrons.

Nucleus of a cell (plural: nuclei). The major *organelle* of *eukaryotes* containing the *chromosomes*.

Oort cloud. A group of comets surrounding the solar system more than a thousand times more populated than the *Kuiper belt*. It is estimated to extend to interstellar distances, perhaps 50,000 AU or more.

Organelle. Structures inside the biological cell that has a given function. Examples are *nuclei, mitochondria* and *chloroplasts*.

Orion Nebula. The largest complex region of interstellar matter known in our galaxy. It is in the Constellation of Orion, some 1,300 *light years* away form the Earth. It is found in the "sword" of Orion surrounding a multiple star system of four hot stars (Theta Orionis, otherwise known as the 'Trapezium'), not older than 100,000 years.

Paleozoic. This is the earliest era of the *Phanerozoic eon.*

Parity violation. The absence of space reflection symmetry. (Usually it refers to the subnuclear world.)

Parsec. A measure of astronomical distance equal to 3.26 light-years.

Phanerozoic. It is the most recent eon of geologic time extending from the *Paleozoic* (570-230 Myr BP) to the *Mesozoic* (230-62 Myr BP) and the *Cenozoic* (62 Myr BP to the present).

Photosphere. The apparent solar surface, where the gas of the atmosphere becomes opaque.

Photosynthesis. An ancient metabolic process, first used by *prokaryotes,* which today produces *hydrocarbons* and liberates oxygen with the source of carbon dioxide and water. In the deep ocean and underground an alternative metabolic process is used (chemosynthesis).

Phylogenetic tree. A graphic representation of the evolutionary history of a group of organisms.
        This may be extended from organisms to species or any higher ordering, such as kingdoms. In molecular biology this concept has been extended to genes as well. The main lesson we may draw form such trees is the path followed by evolution.

Phylogeny. The evolutionary history of an organism, or the group to which it belongs.

Plane-polarized light. Light in which the typical electromagnetic vibrations are rectilinear, parallel to a plane ("plane of polarization") and transverse to the direction of travel.

Plate tectonics. The standard theory, with compelling experimental support, which maintains that the Earth is composed of thick plates floating on submerged viscous material. Each plate moves slowly: its leading edge sinks slowly at regions called subduction zones, while the rear of the plate produces an observed phenomenon called *sea-floor spreading,* where phenomena relevant to the origin of life may occur such as *hydrothermal vents.*

Progenote. A term introduced by the American evolutionist Carl Woese to denote the earliest common ancestor of all living organisms (also called *'cenancestor').*

Prokaryotes. Unicellular organisms that lack a nuclear envelope around their genetic material. Normally they are smaller than nucleated cells *(eukaryotes).* Well known examples are bacteria. All prokaryotes are encompassed in two *domains*: Bacteria and *Archaea.*

Protein. An organic compound, which is an essential biomolecule of all living organisms. Its elements are: hydrogen, carbon, oxygen, nitrogen and sulfur. It is made up of a series of *amino acids.* (A medium-size protein may contain 600 amino acids.)

Proterozoic. In geologic time it is an era that ranges form the *Archean* era (4.5-2.5 Gyr BP) to the *Phanerozoic Eon* (570 Myr BP to the present).

Purine. One of two categories of nitrogen-containing ring compounds found in *DNA* or *RNA.* (Examples are guanine and adenine.); cf., also *pyrimidine.*

Pyrimidine. One of two categories of nitrogen-containing ring compounds found in *DNA* or *RNA.* (Examples are cytosine, thymine and uracil.)

Pyrolysis. Chemical decomposition occurring as a result of high temperature.

Quantum chemistry. A fundamental approach to chemistry based on a mathematical physical theory dealing with the mechanics of atomic systems.

Quartz. A *mineral* whose composition is silica SiO_2.

Reason. A faculty contrasted with experience, passion or *faith.*

Red bed. It is a type of sedimentary rock whose color is due to its ferric oxide *minerals.*

Red giant. A late stage in *stellar evolution* in which the star increases in size and undergoes a change in surface temperature, responsible for its red color.

Refractory. Material that vaporizes only at high temperatures.

Revelation. In this text it is used as a process by which communication of truth by God takes place. Christian philosophers have distinguished between 'truths of reason' and 'truths of revelation'. (cf., *Natural theology.*)

Ribosome. One of many small cellular bodies in which protein synthesis is carried out. It consists of *RNA* and proteins.

Rock. A natural aggregate of one or more *minerals* and sometimes non crystalline substances.

RNA (Ribonucleic acid). Substance present in all organisms in three main forms: *messenger RNA,* ribosomal RNA (cf., *ribosome),* and transfer RNA (used in the last stage of *translation* during protein synthesis). Their general function is related with the *translation* of the genetic message from *DNA* to *proteins*.

Seafloor spreading. The process by which new oceanic crust forms at oceanic ridges by the upsurging of *magma*. During the process of *plate tectonics* the space between separating plates is filled with *magma*.

Sequence. A linear order of the monomers in a biopolymer. In general knowing the sequence is helpful in knowing the function of the molecule.

Shale. Mainly clay that has hardened into *rock*.

Silicate is a *mineral*, common in terrestrial *rocks*, containing silicon and oxygen.

SNC meteorites. These meteorites are included in a class of *basaltic* meteorites, which are of Martian origin.

Solar nebula. Disk of gas and dust around the early Sun that gave rise to the planets and small bodies of the solar system.

Spectral type of a star. This concept refers to a classification of stars according to the appearance of its spectrum.

Spectroscopy. Observational technique that breaks light from an object into discrete wavelength bins for subsequent physical analysis.

Stellar evolution. Stars constantly change since their condensation from the interstellar medium. Eventually they exhaust their nuclear fuel and the *supernova* stage may occur. At this stage *elements* synthesized in its interior are expelled, enriching the interstellar medium, out of which new generations of stars will be born.

Steranes. These are *biomarkers* (i.e., chemical compounds that may be used in dating fossils). Steranes have been found in oils and sediments with ages predating animal fossils.

Steroids. A group of organic compounds chemically similar to *lipids*. They take part in the membrane of eukaryotes and also have different functions. An example is vitamin D.

Stromatolite. A geological feature consisting of a stratified rock formation, which is essentially the fossil remains of bacterial mats. The bacteria that gave rise to these formations were mainly *cyanobacteria*. Similar mat-building communities can develop analogous structures of various shapes and sizes in the world today.

Sulfide. Inorganic compound of sulfur with more *elements* that tend to lose electrons and form positive *ions*.

Supernova. A late stage in stellar evolution. An exploding star that has exhausted its nuclear fuel. It enriches the *interstellar dust*. Its remnant is a *neutron star*.

Symbiosis.   The living together of two (or more) species with presumed mutual benefit for the partners.

Taxonomy. The study of the theory, practice and rules of classification of living or extinct organisms into groups, according to a given set of relationships.

Terrestrial planets. One of the inner rocky planets (Mercury, Venus, Earth, Mars). Due to the similar characteristics of the Moon it is convenient to include it in this group.

Tidal heating. In the case of the vicinity of a satellite to a giant planet heating may be produced due to repeated stressing arising from orbital motion in the planetary field. This source of heating may be enhanced due to the radioactive dacay of nuclei, a phenomenon known as radiogenic heating.

Tissue.   In multicellular organisms, a form of cell organization into cooperative assemblies.

Transcription. A cellular process in which the *DNA* information is transferred to *RNA (messenger RNA),* as a first step in *protein* synthesis.

Translation. A cellular process in which the genetic information is translated into *proteins,* according to the *genetic code.*

Triploblasts. Animals arising from an embryo consisting of three cell layers.

Unsaturated. Term referring to chemical compounds in which the force of attraction holds the atoms together by sharing two pairs of electrons ('double bond') or three pairs ('triple bond').

Viking. US spacecrafts which were able to soft land two sterilized vehicles on opposite hemispheres of Mars in 1976. The mission attempted to detect signs of life amongst the many tasks that were programmed for them.

Volatile. A substance that has a low boiling temperature.

Voyager 1 and 2. US spacecrafts that explored the outer solar system. Although the two Voyagers were planned to depart the same year, they were programmed to explore different aspects of the solar system.

Weak interaction.   Part of a subnuclear interaction, which is five orders of magnitude smaller than its partner, the electromagnetic interaction. After the Glashow-Salam-Weinberg theory, we can think of both partners (the weak interaction and the electromagnetic interaction) as part of a unified 'electroweak' subnuclear interaction.

Zodiacal dust cloud. A tenuous flat cloud of small *silicate* dust particles in the inner solar system.

Supplementary Reading

PREFACE

Ward, P and Brownlee, D. (2000) *Rare Earth,* Copernicus/Springer, New York.

INTRODUCTION

Clarke, A.C. (1993) *By space possessed,* Victor Gollancz, London.

Conway-Morris, S. (1998) *The crucible of creation,* Oxford University Press, London.

Crick, F. (1981) *Life Itself,* MacDonald & Co, London.

Davies, Paul (1998) *The Fifth Miracle The search for the origin of life,* Allan Lane The Penguin Press, London.

Dennett, D. C. (1995) *Darwin's dangerous idea Evolution and the Meanings of Life,* Penguin, London.

Fischer, E. P. and Lipson. W.W. (1988) *Thinking about Science (The life of Max Delbruck),* Norton, New York.

Mayr, E. (1991) *One Long Argument Charles Darwin and the Genesis of Modern Evolutionary Thought,* Penguin Books, London.

Rizzotti, M. (1996) *Defining Life The central problem of theoretical biology,* The University of Padova, Padua.

Schrodinger, E. (1967) *What is life?,* Cambridge University Press, London.

CHAPTER 1: FROM COSMIC TO CHEMICAL EVOLUTION

Cairns-Smith, A.G. (1985) *Seven Clues to the Origin of Life,* Cambridge University Press, London.

Dawkins, Richard (1989) *The selfish gene,* Oxford University Press, London.

Dawkins, Richard (1983) *The extended phenotype,* Oxford University Press, London.

Dawkins, Richard (1988) *The blind watchmaker,* Penguin Books, London.

De Duve, C. (1995) *Vital Dust.Life as a Cosmic Imperative,* Basic Books, New York.

Delsemme, Armand (1998) *Our Cosmic Origins From the Big Bang to the emergence of life and intelligence,* Cambridge University Press,London.

Dyson, F. (1985) *Origins of life,* Cambridge University Press, London.

Eigen, M. and Winkler-Oswatitsch, R. (1992) *Steps towards life. A perspective on evolution.* Oxford, Oxford University Press, London.

Cosmological evolution

Coyne, G., Giorello, G. and Sindone, E. (eds.), (1997) *La Favola dell'Universo.* PIEMME, Como.

Novikov, I.D. (1983) *Evolution of the universe,* Cambridge University Press, London.

Sagan, C. (1980) *Cosmos,* Random House, New York.

Weinberg, S. (1977) *The first three minutes,* Fontana/Collins, London.

CHAPTER 2: FROM CHEMICAL TO PREBIOTIC EVOLUTION

Oro, J. and Cosmovici, C. (1997) Comets and life on the primitive Earth, in C.B. Cosmovici, S. Bowyer and D. Werthimer (eds.), *Astronomical and Biochemical Origins and the Search for Life in the Universe,* Editrice Compositore, Bologna, pp. 97-120.

Owen, T.C. (1997) Mars: Was there an Ancient Eden, in C.B. Cosmovici, S. Bowyer and D. Werthimer, *Astronomical and Biochemical Origins and the Search for Life in the Universe,* Editrice Compositore, Bologna, pp. 203-218.

Winnewisser, G. (1997) Interstellar molecules of prebiotic interest, in C.B. Cosmovici, S. Bowyer and D. Werthimer (eds.), *Astronomical and Biochemical Origins and the Search for Life in the Universe,* Editrice Compositore, Bologna, pp. 5-22.

CHAPTER 3: SOURCES FOR LIFE'S ORIGIN: A SEARCH FOR BIOGENIC ELEMENTS

Chadha, M.S. and Phondke, Bal (1994) *Life in the universe,* Publications and Information Directorate, New Delhi.

Feinberg, G. and Shapiro, R. (1980) *Life beyond Earth: The intelligent Earthling's Guide to Life in the Universe,* William Morrow and Co., New York.

MacDonald, I.R. and Fisher, C. (1996) *Life without light,* National Geographic Magazine, October, pp. 87-97.

McSween, H.Y and Stolper, E.M. (1980) Basaltic meteorites, *Scientific American* **242**, Number 6, pp. 44-53.

Prieur, D., Erauso, G. and Jeanthon, C. (1995) *Hyperthermophilic life at deep-sea hydrothermal vents, Planet Space Sci.* **43**, 115-122.

Raulin, F. (1994) *La vie dans le cosmos,* Flammarion, Paris.

Shapiro, R. (1994) *L'origine de la vie,* Flammarion, Paris.

CHAPTER 4: FROM PREBIOTIC EVOLUTION TO SINGLE CELLS

Bertola, F., Calvani, M. and Curi U. (eds.), (1994) *Origini: l'universo, la vita, l'intelligenza.* Il Poligrafo, Padua.

Brack, A. and Raulin, F. (1991) *L'evolution chemique et les origines de la vie,* Masson,Paris.

Greenberg, J.M., Mendoza-Gomez, C.X. and Pirronello, V. (eds.), (1993) *The Chemistry of Life's Origins,* Kluwer Academic Publishers, Dordrecht, The Netherlands.

Oro, J., Miller, S.L. and Lazcano, A. (1990) *The origin and early evolution of life on Earth,* Ann. Rev. Earth Planet. Sci. **18**, 317-356.

Poglazov, B.F., Kurganov, B.I., Kritsky, M.S. and Gladlin, K.L. (eds.), (1995) *Evolutionary Biochemistry and Related Areas of Physicochemical Biology.* Bach Institute of Biochemistry and ANKO, Moscow.

CHAPTER 5: FROM THE AGE OF THE PROKARYOTES TO THE ORIGIN OF EUKARYOTES

Attenborough, D. (1981) *Life on Earth,* Fontana, London.

Margulis, L. and Fester, R. (eds.), (1991) *Symbiosis as a Source of Evolutionary Innovation,* The MIT Press, Cambridge, Mass.

Margulis, L. and Sagan, D. (1987) *Microcosm,* Allen & Unwin, London.

Schopf, J. William (1999) *Cradle of Life,* Princeton University Press, Princeton.

CHAPTER 6: EUKARYOGENESIS AND THE EVOLUTION OF INTELLIGENT BEHAVIOR

Bendall, GS, ed. (1983) *Evolution from molecules to men,* Cambridge University Press, London.

Darwin, C. (1859) *The origin of species by means of natural selection or the preservation of favoured races in the struggle for life,* John Murray, London.

Gould, S.J. (1991) *Wonderful Life The Burgess Shale and the Nature of History,* Penguin Books, London.

Maynard Smith J. (ed.), (1982) *Evolution now,* Nature, London.

Maynard Smith, J. (1993) *The theory of evolution.* Cambridge University Press, London.

Patterson, C. (1978) *Evolution,* British Museum (Natural History), London.

CHAPTER 7: ON THE POSSIBILITY OF BIOLOGICAL EVOLUTION ON MARS

Carle, G.C., Schwartz, D.E. and Huntington, J.L. (eds.), (1992) *Exobiology in Solar System Exploration.* NASA SP 512 .

Reviews of the Viking Missions (1976) *Science* **193**, 759-815 and *Science* **194**, 1274-1353.

CHAPTER 8: ON THE POSSIBILITY OF BIOLOGICAL EVOLUTION ON EUROPA

Bertola, F. (1992) *Da Galileo alle Stelle,* Biblos Edizioni, Padua. [This book was published during the celebrations at the University of Padua to commemorate the fourth century of the nomination of Galileo Galilei as a lecturer of Mathematics in that university, where he spent 18 years of very productive work.]

Johnson, T. V. (1995) The Galileo Mission, *Scientific American.* December, p. 35.

Europa Ocean Conference at San Juan Capistrano Research Institute. 12-14 November, 1996. Book of summaries (pp. 90). Available from San Juan Capistrano Research Institute. 31872 Camino Capistrano. San Juan Capistrano, CA 92675. Fax: (714) 240-0482. E.mail: swallows@sji.org.

CHAPTER 9: ON THE POSSIBILITY OF CHEMICAL EVOLUTION ON TITAN

Gautier, D. (1997) Prebiotics on Titan: from available to expected measurements, in C.B. Cosmovici, S. Bowyer and D. Werthimer (eds.), *Astronomical and Biochemical Origins and the Search for Life in the Universe,* Editrice Compositore, Bologna, pp. 219-226.

Coustenis, A., Lellouch, E., Combes, M., McKay, C. P. and Maillard, J.-P. (1997) Titan's atmosphere and surface from infrared spectroscopy and imagery, in C.B. Cosmovici, S. Bowyer and D. Werthimer (eds.), *Astronomical and Biochemical Origins and the Search for Life in the Universe,* Editrice Compositore, Bologna. pp. 227-234.

CHAPTER 10: HOW DIFFERENT WOULD LIFE BE ELSEWHERE?

Boss, A. P. (1996) Extrasolar Planets, *Physics Today,* September, pp. 32-38.

Schneider, J. (1996) *Strategies for the search of life in the universe,* in Chela-Flores, J. and Raulin, F. (eds.), *Chemical Evolution: Physics of the Origin and Evolution of Life,* Kluwer Academic Publishers, Dordrecht, The Netherlands, pp. 73-84.

CHAPTER 11. THE SEARCH FOR THE EVOLUTION OF INTELLIGENT BEHAVIOR IN OTHER WORLDS

Drake, F. and Sobel (1992) *Is there anyone out there? The scientific search for Extraterrestrial Intelligence,* D. Delacorte Press, New York.

Heidmann, Jean (1996) *Intelligences extra-terrestres.* Editions Odile Jacob: Paris.

Hoover, R.B. (ed.), (1997) *Instruments, Methods and Missions for Investigation of Extraterrestrial Microorganisms,* Proc. SPIE, **3111.**

Hoover, R.B. (ed.) (1998) *Instruments, Methods and Missions for Astrobiology,* Proc. SPIE, **3441.**

Lemarchand, G.A. (1992) *El Llamado de las Estrellas,* Lugar Cientifico, Buenos Aires.

Raulin, F., Raulin-Cerceu, F. and Schneider, J. (1997) *La Bioastronomie,* Presses Universitaires de France, Paris.

Sindoni, Elio (1997) *Esistono gli Extraterrestri?,* Il Saggiatore, Milano.

Zuckerman, B. and Hart, M.H. (1995) *Extraterrestrials. Where are they?* (2nd. ed.), Cambridge University Press, London.

CHAPTER 12: IS THE EVOLUTION OF INTELLIGENT BEHAVIOR UNIVERSAL?

Wilson, E. O. (1998) *Consilience The Unity of knowledge*, Alfred A. Knopf, New York.

Deacon, T. W. (1997) *The Symbolic Species The Co-evolution of Language and the Brain*, W. W. Norton & Co., New York.

CHAPTER 13: DEEPER IMPLICATIONS OF THE SEARCH FOR EXTRATERRESTRIAL LIFE

Colombo, R., Giorello, G. and Sindoni, E. (eds.), (1999) *Origine della Vita Intelligente nell'Universo (Origin of intelligent life in the universe)*, Edizioni New Press, Como.

Martini, C. M. (1999) *X Cattedra dei non credenti: Orizzonti e limiti della scienza*, Raffaello Cortina Editore, Milano.

Russell, R.J., Murphy, N. and Isham, C.J. (eds.), (1996) *Quantum Cosmology and the Laws of Nature. Scientific Perspectives on Divine Action*, (2nd ed.), Vatican Observatory Foundation, Vatican City State.

Russell, R.J., Murphy, N. and Peacocke, A.R. (eds.), (1995) *Chaos and Complexity. Scientific Perspectives on Divine Action*, Vatican Observatory Foundation, Vatican City State.

Russell, R.J., Stoeger, W. R. and Ayala, F. J. (eds.), (1998) *Evolutionary and Molecular Biology: Scientific Perspectives on Divine Action*, Vatican Observatory and the Center for Theology and the Natural Sciences (a joint publication),Vatican City State/Berkeley, California.

Tough, Tough, Allen, ed. (2000) *When SETI Succeeds:The impact of high information content*, Foundation for the Future: Washington.

CHAPTER 14: PHILOSOPHIC IMPLICATIONS OF THE SEARCH FOR EXTRATERRESTRIAL CIVILIZATIONS

Colombo, R., Giorello, G. and Sindoni, E. (1999) *Origine della Vita Intelligente nell'Universo (Origin of intelligent life in the universe)*. dizioni New Press, Como.

Eligio, Padre, Rigamonti, G. and Sindoni, E. (1997) *Scienza, Filosofia e Teologia di Fronte alla Nascita dell'Universo (Reflections on the Birth of the Universe: Science, Philosophy and Theology)*. New Press, Como.

Davies, Paul (1994) *Siamo Soli? Implicazioni filosofiche della scoperta della vita extraterrestre*, Editori Laterza, Roma.

Russell, R.J., Stoeger (SJ), W.R., and Coyne (SJ), G.V. (eds.), (1995) *Physics, Philosophy and Theology. A common quest for understanding,* (2nd. ed.) Vatican Observatory Foundation: Vatican City State.

Original version (in French) of *The Holy Father Message to the Pontifical Academy* (1997), (the message itself was delivered on 22 October 1996), *Commentari* **4,** No. 3, Part I, The origin and early evolution of life, Pontificia Academia Scientiarvm, Vatican City, pp. 15-20.

CHAPTER 15: BACK TO THE BEGINNING OF ASTROBIOLOGY

Bertola, F. (1995) *Imago Mundi, la rappresentazione del cosmo attraverso i secoli,* Biblos, Cittadella PD, Italy.

Chadha, M. S. (2001) *Reminiscences-Pont-a-Mousson-1970 to Trieste-2000,* in J. Chela-Flores, T. Owen, and F. Raulin (eds.) *The First Steps of Life in the Universe.* Proceedings of the Sixth Trieste Conference on Chemical Evolution. Trieste, Italy, 18-22 September. Kluwer Academic Publishers: Dordrecht, The Netherlands. (To be published.)

Davies, P. (1998) *The Fifth Miracle The search for the origin of life,* Allan Lane The Penguin Press: London.

Desmond, A. and Moore, J. (1991) *Darwin,* M. Joseph: London .

Moorhead, A. (1971) *Darwin and the Beagle,* Penguin: London.

Ricci, S. (2000) *Giordano Bruno nell'Europa del Cinquecento,* Salerno Editrice, Roma.

Sagan, C. (1980) *Cosmos,* Random House: New York.

In addition, we recommend the exhibition catalogue:

Giordano Bruno 1548-1600. Mostra Storico Documentaria (Roma, Biblioteca Casanatense 7 Giugno-30 Settembre 2000). L. S. Olschki Editore. Biblioteca di Bibliografia Italiana 164: Citta' di Castello (2000).

GLOSSARY

We have listed a few valuable dictionaries that will allow the reader a deeper understanding of the subject matter of the present text. We hope that the following works will be of some help in complementing our limited glossary.

Theology

Hinneells, J.R. (ed.), (1984) *The Penguin dictionary of Religions*, Penguin Books, London.

Livingstone, E.A. (ed.), (1977) *The Concise Oxford dictionary of the Christian Church*, Oxford University Press, London.

Philosophy

Flew, A. (ed.), (1979) *A dictionary of philosophy*, Pan Books, London.

Rovatti, P.A. (ed.), (1990) *Dizionario Bompiani dei Filosofi Contemporanei*, Bompiani, Milano.

INDICES

List of abbreviations

Å: Ångstrom, unit of distance corresponding to 10^{-8} cm.

b-Pic: Beta-Pictoris.

BIF : Banded iron formation.

DNA: deoxyribonucleic acid.

ESA: European Space Agency.

GR: General Relativity.

Gyr BP: gigayear (10^9) years before the present.

HIPPARCOS: High Precision Parallax Collecting Satellite.

HGT: horizontal gene transfer.

H_0: Hubble constant.

HR: Hertzsprung-Russell, names associated with the famous graphic display of the main-sequence stars.

HST: Hubble Space Telescope.

IDEA: Instituto de Estudios Avanzados, Universidad Simon Bolivar.

IDP: Interplanetary dust particle.

IRAS: Infrared Astronomical Satellite.

ICTP, The Abdus Salam International Centre for Theoretical Physics, formerly called simply International Centre for Theoretical Physics.

ISSOL: International Society for the Study of the Origin of Life.

JPL: Jet Propulsion Laboratories.

MGS: Mars Global Surveyor.

MSSS: Malin Space Science Systems.

Myr BP: million years before the present.

NASA: National Aeronautics and Space Administration.

NGST: Next Generation Space Telescope.

NICMOS: Near Infrared Camera and Multi-Object Spectrometer of the HST.

NSF: National Science Foundation.

PAL: present atmospheric level.

pc: parsec.

ppm: parts per million.

R: the scale factor of cosmology.

RNA: ribonucleic acid.

SETI: acronym for search for extraterrestrial intelligence.

SNC: acronym for a specific kind of meteorite. The initials stand for the locations where these meteorites were first retrieved (shergottites, nackhlites and chassignites, respectively) .

TPF: Terrestrial Planet Finder.

UG: Universal Grammar.

USB: Universidad Simon Bolivar, whose campus is in Caracas, Venezuela.

UV: ultraviolet.

yr BP: years before the present.

Subject Index

(Terms with an asterisk (*) are explained in the Glossary)

Name Index

PROCEEDINGS OF RELATED EVENTS IN ASTROBIOLOGY
CO-EDITED BY THE AUTHOR

Ponnamperuma, C. and Chela-Flores, J. (eds.), (1993) *Chemical Evolution: Origin of Life,* A. Deepak Publishing, Hampton, Virginia, USA.

Chela-Flores, J., Chadha, M., Negron-Mendoza, A. and Oshima, T. (eds.), (1995) *Chemical Evolution: Self-Organization of the Macromolecules of Life,* A. Deepak Publishing, Hampton, Virginia, USA.

Ponnamperuma, C. and Chela-Flores, J. (eds.), (1995) *Chemical Evolution: The Structure and Model of the First Cell,* Kluwer Academic Publishers, Dordrecht.

Chela-Flores, J. and Raulin, F. (eds.), (1996) *Chemical Evolution: Physics of the Origin and Evolution of Life,* Kluwer Academic Publishers, Dordrecht.

Chela-Flores, J. and Raulin, F. (eds.), (1998) *Chemical Evolution: Exobiology. Matter, Energy, and Information in the Origin and Evolution of Life in the Universe,* Kluwer Academic Publishers, Dordrecht.

Chela-Flores, J., Lemarchand, G. A. and Oro, J. (eds.), (2000) *Astrobiology From the Big Bang to Civilization,* Kluwer Academic Publishers, Dordrecht.

Chela-Flores, J., Owen, Tobias and Raulin, F. (eds.), (2001). *The First Steps of Life in the Universe.* Kluwer Academic Publishers: Dordrecht, The Netherlands. (In preparation.)

Cellular Origin and Life in Extreme Habitats

KLUWER ACADEMIC PUBLISHERS – DORDRECHT / BOSTON / LONDON